The
Electrician's Bible

The Electrician's Bible

A HomeOwner's Bible

PETER JONES

DOUBLEDAY & COMPANY, INC., GARDEN CITY, NEW YORK

Library of Congress Cataloging in Publication Data

Jones, Peter, 1934–
   The electrician's bible.

   1. Electric engineering—Amateurs' manuals.
2. Electric wiring, Interior—Amateurs' manuals.
3. Household appliances, Electric—Maintenance and
repair—Amateurs' manuals. I. Title.
TK9901.J68   621.319′24
ISBN: 0-385-17345-8
Library of Congress Catalog Card Number 80–2552

Contents

The Electrician's Bible

CHAPTER ONE

The Fundamentals of Electricity

You do not have to know anything about the theory of electricity to successfully work with electrical systems. You only need to understand it if you plan to be a teacher or go into the electronics business. On the other hand, the theory behind electricity is excellent fodder for impressing and/or boring your friends and neighbors. So, because this book is an effort to provide all the information you need to have in order to accomplish your own home electrical work, this first chapter is dedicated to some of the more basic theoretical explanations of electricity, what it is, and what it does.

THE ELECTRICAL THEORY

Scientists laboring long and hard during the past century have concluded that electricity is an invisible force that operates between the atoms that make up all matter. You can't feel it, see it, smell it, hear it, or touch it, but electricity can be used nevertheless to produce heat, light, sound, and motion. Since scientists like to explain every phenomenon that occurs in the human experience, their theory is broken down like this:

Matter is anything that takes up space and has mass (weight). Every substance, whether it is a solid, a gas, or a liquid, is made up of atoms, which in turn are composed of neutrons, protons, and electrons (plus some other tiny items that do not really have to be considered in a discussion of electricity).

Neutrons are, electrically speaking, neutral. They have neither a positive nor a negative electrical charge.

Electrons have been assigned the characteristic of having a negative polarity.

Protons have been given the characteristic of a positive polarity.

Positive and negative polarities are opposites, just as the north and south poles of a magnet are opposite. And, like the poles of a magnet, similar electrical charges tend to repel each other, while opposite electrical charges tend to attract each other.

THE ATOMIC THEORY

Actually, the theory is that all of the neutrons and protons group together in the center of an atom and form a stable nucleus. At the same time, all of the electrons in the atom orbit around the nucleus, the same way planets orbit around the sun. All that concentrated positive electrical force pulls at each of the orbiting electrons, trying to suck them toward the center of the atom. But the momentum of the electrons in orbit balances the pull of the nucleus, so, intrepid little fellows that they are, the electrons keep right on whiz-banging around the neutrons and protons and never attach themselves to the nucleus.

As long as there are an equal number of electrons and protons in a given atom, their negative and positive charges cancel each other out and there is no evidence of electricity. Thus, the atom is said to be neutral. The *potential* for electricity is there, however, and if some of the electrons can be separated from their orbits around the nucleus, the electrical balance of the atom will be changed, and with that change comes an electrical charge.

The electrons in an atom surround the positive (proton, neutron) nucleus in a series of orbital rings known as shells. The electrons closest to the nucleus are in the unenviable position of fighting a stronger pull from the protons than the ones in the outer shells, so they tend to remain in their orbits and are considered stable. The electrons in the outer shell, however, have less pull on them, and are consequently a little shakier. In fact, in materials such as copper, the electrons in the outer shell (known as the valence ring) are downright untrustworthy.

Copper atoms all have 29 protons in their nucleus and 29 electrons orbiting in four shells (Fig. 1). But there is only one electron in the outer valence ring, and it is so far away from the nucleus that there is almost no pull on it whatsoever. This means that the valence electron very easily can go philandering off to join some other atoms. In fact, theorists consider the valence electron in a copper atom to be so flirtatious and apt to go from atom to atom that it is called a "free electron." And, it is the free electrons moving from atom to atom that produce electrical current in copper wire or any other electrical conductor.

Current can only be caused when a valence electron leaves the atom it belongs to and attaches itself to another atom, at which point both atoms are out of balance. The original atom has one too many protons and the second atom has one too many electrons, so both of them are considered "unbalanced" and are called *ions*.

IONS

An ion can be either negatively or positively charged. If the atom is a negative ion, it has more electrons than protons and, therefore, possesses a negative electrical charge. If the atom is a positive ion, it has more protons than electrons and subsequently carries a positive charge. So, if you want to create electricity, you have to push the valence electrons from atom to atom in a given substance.

STATIC ELECTRICITY

The simplest way of producing the force that will push the valence electrons through a substance is with friction. You can rub a balloon against your wool sweater, for example. Right away the valence electrons in the wool atoms will jump over to play with the atoms that make up the surface of the balloon. That makes the surface of the balloon negatively charged (because its atoms now have an excess of electrons) and leaves the surface of the sweater positively charged (because its atoms are left with more protons than electrons). As a result, because opposites attract, the balloon will cling to the sweater. It will also cling to a wall because the negative charge of its ions will naturally repel the electrons from the surface of the wall, making the surface atoms positively charged.

What holds the balloon in place is known as *static electricity*. Static electricity means simply that it is stationary. It just sits there and doesn't go anywhere, nor does it produce a continuous flow of electrons. To make the electrons truly flow, you have to put continuous pressure on them.

Static electricity has some useful applications, such as drawing the negatively charged particles of abrasive against a piece of paper to create some types of sandpaper, but the uses are definitely limited when compared to what can be done with a constant flow of electrons through a conductor.

CONDUCTORS

A conductor is any material consisting of atoms having valence electrons that can easily be moved by an electromotive force, known as *volt-*

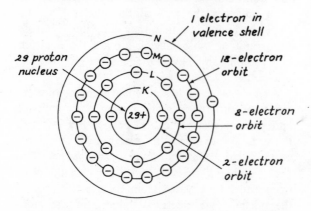

Fig. 1. In theory, this is what a copper atom looks like.

age. Of all the substances known to man, the most practical ones to use as electrical conductors are metals; among these, silver, copper, gold, aluminum, and iron have the most useful atomic structures. All of them are made up of atoms that contain relatively unstable valence rings with one or more free electrons that can be pushed through the substance from atom to atom.

INSULATORS

The opposite of a conductor is an insulator. Insulators are tough guys. They have atoms with valence electrons that do not migrate freely, and therefore they do not conduct electricity very well. Insulators such as glass, plastic, rubber, mica, and air are therefore used to contain electricity. For example, copper electrical wires with their valence electrons dancing all over the place can be rendered harmless to the touch by wrapping them in plastic or rubber. This is because the valence electrons in such insulating materials are very grudging about leaving the atoms they belong to.

ELECTROMOTIVE FORCE (VOLTAGE)

Since electrons are marvelously tiny, most practical applications require that the combined charge of many billions of them be moved through a conductor at high speeds. If you leave the electrons in a copper wire alone, they will drift through the wire at a rate of less than one inch per second, and will not produce very much electrical current in the process. But if an electromotive force is used to push them, they will travel at the speed of light (186,000 miles per hour) and become a continuous flow, which is generally referred to as electrical current. Again, to get an electrical current, you have to have a source of electromotive force, or voltage.

Voltage is created in lots of ways. Chemical reaction will cause it, for example, and so will magnetism or a machine called a generator. But unless this source is connected to a conductor, such as copper wire, there will be no electrons for the voltage to push and nothing electrical will happen. When a conductor (usually a wire) *is* attached to any source of voltage, it must begin at the source and end at the source, i.e., *it must*

form a complete, unbroken circle. If you connect only one end of a copper wire to one terminal of a storage battery, electrons will move through the wire but will have nowhere to go when they get to the unattached end (remember, air is an insulator, not a conductor). As soon as the free end of the wire is attached to the other terminal of the battery, however, you have a different ball game. Now, the electrons can do their thing.

Any source of voltage moving electrons through a conductor in a continuous flow is doing a certain amount of work. That amount of work is reckoned in *volts,* which is usually symbolized by the capital letter V. But just to make things hard to remember, the capital letter E is also used (for Electromotive force). One volt is the amount of electromotive force required to do 0.7376 foot-pounds of work in moving a charge of 6.25×10^{18} electrons between two points. If you are good at remembering irrelevant information and still want to impress your friends, try this: An electrical charge of 6.25×10^{18} electrons is called one *coulomb,* and 0.7376 foot-pounds is equal to the practical metric unit of work or energy known as one *joule.* Therefore, one volt equals one joule of work per coulomb of charge. If you aren't inclined to spout incomprehensible definitions over the breakfast table, just remember that voltage is nothing more than a force that pushes electrons through a wire. Volts don't go anywhere, by the way. They just stay wherever they are produced and push. You will never find one in any conductor.

ELECTRICAL CIRCUITS

Inside a battery there are two plates made of dissimilar metals and submerged in chemicals. Chemical action between the plates pulls the electrons from the positive plate to the negative plate, which is attached to the negative terminal on the outside of the battery (it has a little — sign printed on or next to it). The negatively charged valence electrons in the negative plate merrily jump from atom to atom in the wire and keep right on moving until ultimately they get back to the positive plate in the battery via the positive terminal (which has a little + sign printed on or next to it). This is the basic electrical circuit.

Every electrical circuit must contain three major components: a source of electromotive force, a conductor, and a circuit load (Fig. 2). The source of voltage can be dry cells, batteries, or generators—anything that can supply enough force to push electrons through a conductor. The conductor can be any one of a variety of materials, but for most practical purposes it is a metal, usually copper or aluminum because they are the cheapest available. The circuit load can be any electrically powered device such as a lamp, a motor, or a household appliance. It is important to remember that every electrical circuit works the same way. If there is an opening anywhere in the conductor, the electrons stop flowing.

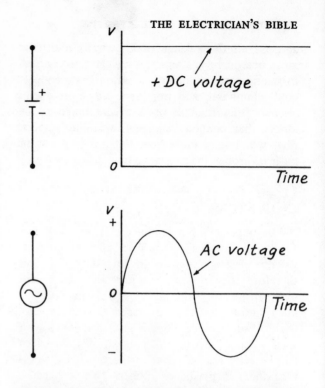

Fig. 3. Steady DC voltage can be produced by a battery (top) having one polarity. A sine-wave AC voltage (bottom) has an alternating polarity. The bottom drawing shows one complete cycle of AC voltage.

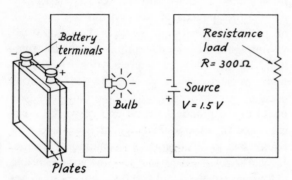

Fig. 2. Every circuit must have a source (the battery), a conductor (the wire), and a load (the light bulb). The funny-looking box on the right is the electrical industry's way of showing the drawing on the left in schematic form.

DIRECT CURRENT (DC)

Electrical current, i.e., a flow of electrons, travels at the speed of light in one of two ways: either *direct current* or *alternating current*. Direct current is a fixed-polarity, single-direction movement of electrons through a conductor. They come zooming out of the negative pole of whatever electromotive force is pushing them, zip through the conductor and the circuit load, and come back via the positive pole. It should be noted, however, that by general convention and agreement, when anyone is discussing an electrical circuit, electricity is analyzed as if it traveled from the positive to the negative pole, which is exactly the opposite of what really happens

(presuming you swallow all this theory about electrons to begin with). So, conventionally speaking, current flows from the positive to the negative poles of the source of electromotive force.

ALTERNATING CURRENT (AC)

Alternating current also flows from the positive to the negative pole (in conventional terms), but it doesn't go in a straight line the way direct current does. Alternating current flows in one direction for a given period of time, then reverses itself for an equal length of time and flows in the opposite direction. It can do this because of the design of the generators that produce it. These machines reverse their polarity every split second so that the positive terminal becomes negative and the negative terminal becomes positive. Since the current is traveling at the speed of light, nothing is interrupted.

Electrical current is symbolized by the capital letter I, but the basic unit for measuring it is the *ampere,* which is symbolized by the capital letter A. Consequently, amperage is often used as a synonym for current and is defined as equal to 6.25×10^{18} electrons passing a single point per second. Voltage does not move through a conductor, but amperes do, and it is the amperes that will sting you when you touch the bare metal of a conductor.

CIRCUIT LOADS

The last major component of a circuit is its load, which can be any device that needs electricity to operate. In effect, a circuit load is in the business of converting electricity into work or energy. For example, a lamp turns electricity into light, a toaster converts it to heat, and a motor uses it to perform mechanical labor. The important fact about circuit loads is that they each offer a certain amount of opposition to the current passing through them. How much opposition depends on what the load is and what materials were used to make it; that opposition is always referred to as *resistance.* Resistance is symbolized by the capital letter *R,* printed in italics. The unit for measuring resistance is the *ohm,* which is symbolized by the Greek letter omega (Ω). An ohm is defined by a ratio of volts to amperes, which brings us to a mathematical formula known as Ohm's Law.

OHM'S LAW

Ohm's Law states that resistance in any circuit is equal to the voltage applied to the circuit divided by the current in the circuit. In other words, ohms equal volts divided by amperes. As a practical electrician working with your house wiring system, you will not have to get involved with Mr. Ohm and his Law very often, if at all, but there is an easy way to cohabitate with it. Simply write it as a fraction, V divided by I times *R,* and put it in a triangle (Fig. 4). To determine any unknown value, cover the appropriate symbol on the triangle and work out the part of the formula that is left. For example, if you know the voltage and resistance, cover the I. Then divide ohms into volts to get the amperage, or current. If you want to find the resistance, cover the *R* and the remaining formula will

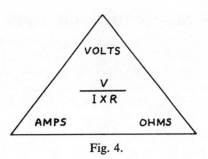

Fig. 4.

tell you to divide amperes into volts. If you need to know the voltage, covering the V on the formula tells you to multiply the amperes times the ohms.

THE POWER FORMULA

There is another useful formula for determining electrical relationships that can also be handled in a triangle. This formula states that power equals voltage times current. The symbol for power is a capital letter *P* printed in italics; the unit of power is the *watt,* which is given the capital letter W. One watt equals the work in joules done in one second by one volt moving one coulomb of electrical charge. In other words, a watt is the total amount of power that flows at a given point, so watts equal amperes times volts.

Again, cover the symbol for whatever value you are looking for and then read the remaining formula (Fig. 5). For example, to find watts, cover the W and the triangle will tell you to multiply $I \times E$, or amps times volts. Or, cover the E and the remaining formula tells you to divide watts by amps.

Because Ohm's Law is applied primarily to direct current circuits and there are few such circuits in the average home, it does not have much bearing on your electrical work. The power formula is another animal because it is primarily applicable to alternating current circuits.

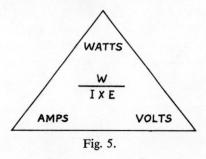

Fig. 5.

FROM THEORY TO PRACTICALITY

For all of the guestimates and terms and formulas that have been devised to explain electricity, there have come some very practical applications which our society uses to generate heat, light, magnetism, and motion. But when all of the theories have been expounded, electricity comes down to one very simple basic principle:

In order for an electrical circuit to function, a conductor (wire) must be attached to both poles of a source of electromotive force. If you have an electrical circuit that is receiving voltage but is not operating whatever appliances you have connected to it, there is a break in the circuit. In other words, the conductor is not forming a complete, unbroken circle from the source of voltage and back, and the circuit is said to be open. That is what home wiring repairs and maintenance are all about: closing the gaps in circuits that provide your house with electrical power.

CHAPTER TWO

Electricity in Your Home

All over the country there are giant (two- and three-story high) alternating current (AC) generators operating around the clock to provide electricity. These generators belong to a network of utility companies and use a variety of fuels including coal, oil, nuclear power, and, cheapest of all, water. The fuel is used to rotate gigantic loops of copper wire (that are wrapped around a metal core) through a magnetic field created by still more loops of wires. This is what generates AC current.

Alternating current generators produce a current that travels in one direction for $\frac{1}{120}$th of a second, then reverses itself for the next $\frac{1}{120}$th of a second. Each direction the current takes is known as an *alternation* (hence the term alternating current). Two complete alternations are called a *cycle,* and the number of cycles completed per second is the *frequency* of the current. The frequency of a current can be anything, depending on the speed of the generator creating it. Throughout Europe the standard frequency is 50 cycles, or *hertz* (Hz). In the United States all current entering homes is 60 hertz (Hz) (Fig. 6),

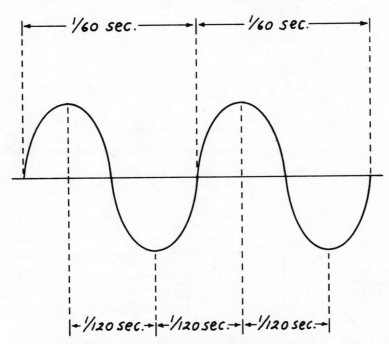

Fig. 6. Two of the 60 cycles of 60-Hz alternating current.

although some industries must have alternating current anywhere from 25 Hz to 400 Hz. For any homeowner in the United States working with his house electrical system, the current encountered will be a single-phase current alternating at 60 times a second.

PHASES

The electricity coming from AC generators is actually not one current, but three. They are all alternating at 60 times a second, and in fact are completely identical. They are timed so they do not interfere with each other and are known as *phases*. Three-phase electricity is not particularly advantageous to households, because home appliances are all built to operate on single-phase AC, but 3-phase does improve the efficiency of many motors used in industry. If a motor is run with single-phase AC, it receives a "push" 120 times a second, i.e., every time the current changes direction. But if it is operating on 3-phase current, the peak voltage of each phase arrives at the motor at a different moment, giving the machine 360 "pushes" per second.

STEPPING UP AND STEPPING DOWN

As soon as voltage has been produced by the utility company generators, the current must be sent hundreds of miles to each of the company's customers. The trouble is, the resistance to current from the conductors it must travel through is very likely to dissipate much of the electricity before it can reach its destination. So, the power companies step up the voltage to as much as 240 kilovolts by pushing it through a *transformer*. Transformers, by the way, are nothing more than coils of different-sized wire; they have no mechanical parts. The power is then able to travel long distances with considerably less loss in voltage. By the time it reaches its destination, however, it is still too high to be used by any home appliances, so it must be reduced by pushing it through a series of *step-down* transformers that decrease the current to something between 30 and 200 amperes and 120 and 240 volts. The last of these step-down transformers is buried in the ground or hanging off a pole right outside your house, in the form of what looks like a 40-gallon garbage can. There may even be several of them on the same pole, each serving different sets of homes in your neighborhood (Fig. 7).

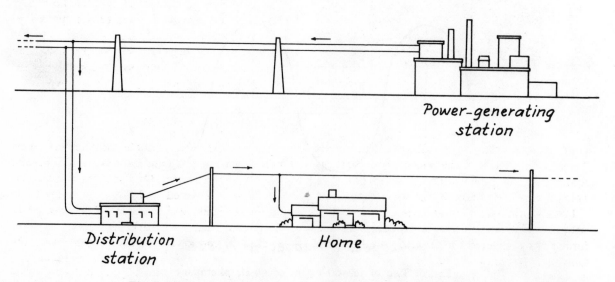

Fig. 7. Where electricity comes from and how it gets to your house.

HOME DELIVERY

Once the power has been reduced to a manageable amount, it is brought into your house via a *feeder cable* which passes through a utility company meter and ends at the *distribution box* inside your home (Fig. 8). You can tell exactly what the feeder cable looks like, and consequently the kind of electrical service you are receiving, simply by reading all the fine print on

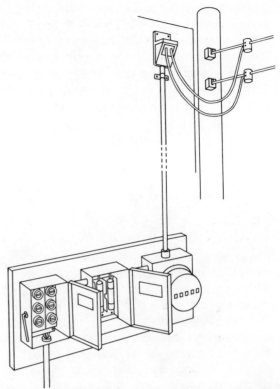

Fig. 9. The electricity arrives at your house and passes through a kilowatt-hour meter to the main current interrupter and ends at your distribution panel (fuse or circuit-breaker box).

Fig. 8. Needless to say, the inside of any distribution box is a maze of wires.

your electric company's kilowatt-hour meter. There is a plate on the meter face that says something like "240 volts, 3 wire," or "110 volts, 2 wire." And therein lies some history.

Until World War II most homes were supplied with a 2-wire, single-phase current from a transformer that delivered the electricity at 110 volts. Only a few prewar homes were supplied by a 3-wire cable that allowed the use of both 110 volts and 220 volts.

During the building boom right after World War II, the utility companies realized that there would be an increased demand for power as more and more home appliances came onto the market, so they began to supply practically all new constructions with single-phase, 3-wire, 120/240-watt circuits.

THREE-WIRE SYSTEMS

The 3-wire feeder cable that enters most homes that are less than forty years old literally consists of three wires. One of them is a neutral wire that is covered with white or near-white plastic insulation and is invariably connected to the ground (i.e., the earth). It is either attached to a metal rod buried in the ground or to the cold-water supply pipe that brings water into the building. The other two wires in the feeder cable are "hot" and are insulated with black or any

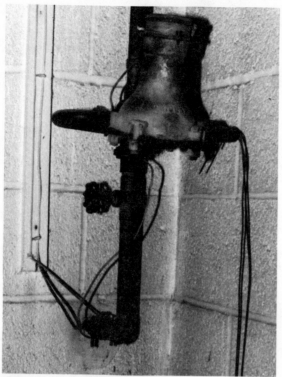

Fig. 10. The bare grounding wires coming down the wall are run directly from the distribution box to the water meter where they are clamped to the incoming water supply pipe.

color other than white or green plastic. Each of the two hot lines carries 120 volts, so if you connect any device to either of the black wires and the white one where it is grounded, you will have 120-volts power. If you connect a device to the two black (and the white) wires, you will receive twice 120 volts, or 240 volts. Consequently, the 3-wire system gives every house owner considerable versatility so far as the electricity that is available to run home appliances; you can use 120 volts for all of the small appliances and lights in the building, but also have 240 volts available for running such large-capacity devices as electric ranges, hot water heaters, clothes dryers, and air conditioners.

There is one hidden possibility so far as the voltage entering your home is concerned: it may not be a true 120 or 240 volts all the time. It may, in fact, be only 104 or 208 volts, or something in between, depending on your locality and the capability of the local utility company to deliver voltage. From the standpoint of operating appliances, you will hardly, if ever, notice the difference. But electrical calculations are often made using 115 and 115/230, or 110 and 110/220 volts, which is why you will still encounter professionals who refer to 110 and 220 volts. For the sake of simplicity, all references to house voltage in this book will use 120 and 120/240 volts.

TOTAL POWER

Three-wire cable commonly serves homes wired to handle 60 amperes or more, but 100 amperes is practically a standard with modern construction, and 150 amperes is considered ideal. A 3-wire 120/240-volt system with 150-amp service provides 36,000 watts of power (240V×150A) to the home, which is sufficient to run just about any appliances you might assemble in your domicile, short of using electricity for central heating or air conditioning. For that, you need 200 amps of service.

THE KILOWATT-HOUR METER

Before you even begin to use any electricity in your home, the feeder cable goes through a utility-company-owned meter which dutifully ticks off every smidgeon of Mr. Edison's "Magic Flooid" that enters your house. Usually the meter is mounted on an outside wall, although it may be in the basement or wherever your service entrance and distribution box (where the fuses are) are placed. The meter itself is a marvel of durability, since it is tamperproof, weathertight, and rarely needs to be replaced. It also belongs to your local utility company and fooling around with it is a criminal offense. But you are allowed to read it, just as the electric company does every month, to see how much electricity you are paying for. If you don't bother to read it, the utility will tell you anyway on its monthly bills. Barring a foul-up with the company computer, it will usually be an accurate reflection of both the amount of electricity you use and the soaring price of electrical power.

There are presently digital readout meters going into service in various parts of the country that have a faceplate that looks like a speedom-

Fig. 11. Most kilowatt-hour meters not only have dials but a considerable amount of data printed on their faces.

can arrive at some reasonably sized numbers to work with when it is making out your monthly bill. Once the quantity of kWh is identified by your meter, it is multiplied by whatever the rate happens to be. The rate can be anything from 4 or 5 cents to 12 or 15 cents a kWh, depending on where you live.

READING YOUR ELECTRIC METER

Most of the kilowatt-hour meters now in use have four or five numbered dials on their face. Each dial has a single pointer that rotates past the numbers 0 to 9. Look at the dials closely. The dial on the far left has a 0 at its top and begins numbering 1 to 9 going left. The second dial is reversed, with the numbers going to the right, as if it were the face of a clock. The next dial is backward again, and the last dial on the far right is identical to the second dial. There is also a little arrow printed on each dial that indicates which direction the pointer is rotating, just in case you forget that the pointers always rotate in the direction of the numbers from 1 up to 9.

When you read a kilowatt-hour meter, start by reading the meter from left to right. At each dial, write down the *lowest* of the two numbers between which the pointer is resting. The *lowest* number. If the pointer is between 2 and 3, even if it is almost touching the 3, the dial is reading 2.

Any one reading is meaningless, of course, unless you have a second reading to compare it with. If you do begin a regular series of readings, you will be able to keep track of the amount of electricity your home is consuming and it may help you to conserve electricity and therefore reduce your monthly electric bill. For example, when the weather gets hot you will become abundantly aware of the fact that you are running your air conditioner around the clock—at a dramatic increase in cost. With a little experimentation in electrical conservation, you will discover that merely by turning off every light or appliance when they are not really needed you can cut your monthly electric bill by a surprising amount.

eter. If you have one of these, you need only read the numbers from left to right. What the numbers represent at any given moment is the amount of kilowatt hours of electricity that have passed through the meter since the time it was installed, so you won't know much about your usage unless you read the meter at least twice. The first time you read it might be on the first day of a new month. You would record the number and then wait a month to read the meter again. At that point you subtract the first reading from the second reading, which will tell you how many kilowatt hours of electricity you have consumed during the intervening month.

A *kilowatt hour,* which is symbolized as kWh, is 1000 watts of electricity consumed in one hour of use. That doesn't mean you are using 1000 watts every hour of the day and night, but if you turn on a 100-watt light bulb and leave it on for ten hours, you will have consumed 100 watts times 10 hours, or 1000 watts of power. The reason for all this higher mathematics is that a single watt is comparatively small, and by multiplying everything by 1000, the electric company

MAIN SERVICE ENTRANCE

Once the feeder cable enters your house, it is immediately connected to a main overcurrent protection device which takes the form of *fuses* or *circuit breakers*. In older homes overcurrent devices are often housed in their own metal box, but in modern home construction the main overload device is usually in the same distribution box that contains the fuses or circuit breakers that protect the branch circuits in the house. Either way, it is the main overcurrent device that controls all of the power entering your home, and by turning it off (if it is a circuit breaker) or removing it (if it is a fuse) you can close off all power coming into the building.

It should be noted that if the main overcurrent device is housed in a separate box, the unit normally cannot be opened until a lever on its side has been pulled. The lever is attached to the fuses inside the box as well as the cabinet door, so you are protected from even going near the fuses before they have been disconnected, and all of the power in the building has been shut off.

THE DISTRIBUTION PANEL

In modern constructions, the feeder cable from the utility company's distribution system ends at the main overcurrent device in the distribution panel. At this point you are getting your full

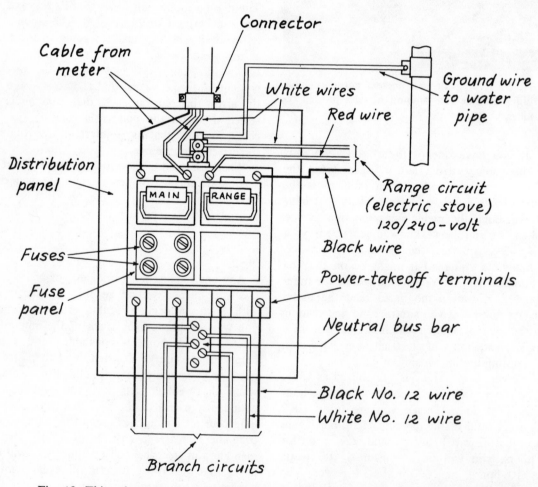

Fig. 12. This schematic of a distribution panel does not contain the bare grounding wires for the sake of clarity. The power-takeoff terminals can be used to connect the panel with an add-on distribution box, as well as for hooking up the branch circuits.

100-amp, 120/240-volt service (or whatever the utility is providing you with). That electricity is now immediately distributed to a bank of fuses or circuit breakers positioned in the panel, each of which controls a different circuit running through the building. The circuits are commonly referred to as branch circuits, and a typical home will have three different types of branch circuits serving it.

Each branch circuit consists of a 2- or 3-wire (and on occasion 4-wire) cable which begins at the distribution panel (Fig. 12). The white, or neutral, wire in each cable is attached to a neutral *bus bar* found usually at the bottom of the panel. The black wire in the cable is connected to one of the fuse or circuit-breaker sockets in the panel face, while the bared grounding wires from all of the cables are brought together to terminals on the neutral bar and reside with the white neutral wires (bare wires deleted from Fig. 12 for clarity).

The result of the wiring configuration in the distribution panels is that a circuit interrupter (fuse or circuit breaker) is stationed between the incoming power and each of the house branch circuits. The purpose of every interrupter is to protect the circuit it controls from overheating and causing a fire.

CURRENT INTERRUPTERS

Whether they are fuses or circuit breakers, all current interrupters are rated for how much electricity they will allow in a circuit. The rating is stated in amperes somewhere on the face of the device. But the fuses or circuit breakers are not just stuck in your distribution board at the whim of some long-gone electrician. Their ratings can be 15, 20, 25, 30, 40, 50, or 60 amperes, depending on the *ampacity* of the 14-, 12-, 10-, and 8-gauge wire used to make up the branch circuits in your home. The wire gauge selected for a given circuit depends on several factors which will be discussed later in Chapter Three.

So long as no excessive current is being drawn through a branch circuit (which is most of the time), the current interrupter does nothing but sit there at the gateway between the incoming power and the beginning of the branch circuit. But let more current than the current interrupter

device is rated for pass through it and the interrupter will immediately "blow" and break the connection between the incoming power and the branch circuit, shutting off all electricity in the circuit it protects.

OVERLOADED CIRCUITS

When electric current flows through a wire, it causes the wire to become hot. As the number of amperes increases, the temperature of the wire increases in proportion to the square of the current. If you double the current, the heat will increase by four times; triple the current and the heat increases nine times. The hotter the wire becomes, the more the likelihood that it may damage the insulation around it and perhaps even cause a fire. Fires, it need not be pointed out, can cause both property damage and the loss of life.

Therefore, it is in everyone's interest to limit the amount of current passing through any conductor to a value that can safely be carried by the size (gauge) and type of wire being used. The method of limiting that amount of current is to install fuses or circuit breakers on each circuit. Thus, if a 15-amp circuit is supplying power to your refrigerator, a toaster, and a coffeemaker, the fuse may allow the refrigerator to turn on and draw the current it needs to operate because it is not demanding more than the 15 amperes that the circuit can safely provide. The circuit may even be able to provide enough electricity to run the toaster at the same time the refrigerator is going. But turn on the coffeemaker while the refrigerator is running and you are making your morning toast and poof!—everything shuts off at once and you are in for a trip to the basement to install a new fuse.

Since people get tired of going to the basement every morning to change a fuse just so they can finish making breakfast, they stop putting 15-amp fuses in the distribution panel and start using 20- or 25-amp fuses. Now they don't have to interrupt the morning ritual anymore, but smoldering away in the walls between the cellar and the kitchen is a branch circuit cable rated at 15 amperes that is being asked to carry 20 amperes. With luck, nothing will ever happen; more likely there will come a day (or night) when that overloaded circuit causes a house fire. Needless to say, overfusing any branch circuit in your home is a very dangerous game to play.

SHORT CIRCUITS

Short circuits can be caused by such things as overloading a circuit, a loose wire connection, or a faulty lamp or appliance. If a wire carrying current should in any way touch another wire, or perhaps the frame of a piece of electrical equipment, the voltage source will immediately produce an infinitely high value of current because it is limited only by the relatively small resistance through its current path. The result is a surge of voltage that often (but *not always*) is enough to cause the circuit interrupter to "blow" or turn off.

Short circuits often occur when the insulation around a power cord becomes frayed, baring the wires inside the cord. The bared wires touch at some point and the current begins to flow between them, forcing the current interrupter to blow, perhaps causing a spark and scaring the user.

Another type of short circuit takes place inside an appliance when a wire works loose from its terminal and touches the housing around it. Again the amperage is increased, again the circuit interrupter blows, but this time anyone touching the housing of the appliance may also get a shock. It is also possible for the inside of the appliance to heat up enough to become really hot and cause a fire.

Still a third type of short circuit takes the form of an electrical arc between a hot wire and some part of a machine that is hot enough to start a fire. This type of short circuit constitutes a shock hazard to anyone holding the appliance, a hazard that could be severe if the person is also in contact with water at the time it occurs.

THE DANGERS OF ELECTRICAL SHOCK

The general notion is that high voltage causes fatal shocks. While that is sometimes true, it is really the amount of current that flows through the body that determines the severity of a shock. If less than one milliampere (one thousandth of an ampere) enters the human body there is noth-

ing more than a mild tickle. Up to 15 mA the experience is unpleasant, but the victim can usually let go of the wire of the appliance causing the shock. More than 15 mA and muscular "freeze" is likely to set in, which prevents the person from freeing himself from the object. And if you can't let go of the voltage source, the shock is probably going to be fatal.

The way to protect yourself against shocks is to be certain your entire house electrical system is properly grounded and that all safety precautions, as set forth in the National Electric Code and your local building codes, have been observed.

FUSES

There are five types of fuses used in house electrical circuits. Essentially, all fuses are a short length of metal ribbon constructed of an alloy having a low melting point. The ribbon is engineered so that it will carry a specified amount of current indefinitely but will melt the moment any current in excess of its rating passes through it. When the ribbon inside a fuse melts, it breaks; when it breaks, it severs the link between the incoming power and the branch circuit it is attached to and opens the circuit.

PLUG FUSES

These are perhaps the most common fuses in America. Their fusible link (the metal ribbon) is contained in a small threaded housing that is screwed into a socket in the distribution panel. There is a thick mica window across the face of the housing so that you can see if the ribbon is melted or not. Plug fuses are rated at 15, 20, 25, and 30 amperes and can be installed in practically any circuit in the house, although the 240-volt circuits, even if they are less than 30 amperes, normally use *cartridge fuses*.

A plug fuse is threaded into its socket by rotating clockwise until it cannot be turned any more by hand pressure. To determine if a plug fuse is blown, look at the fusible ribbon: if it is merely broken, the fuse has blown because of an overload on the circuit; if the mica is smudged or blackened, you may not be able to see the metal

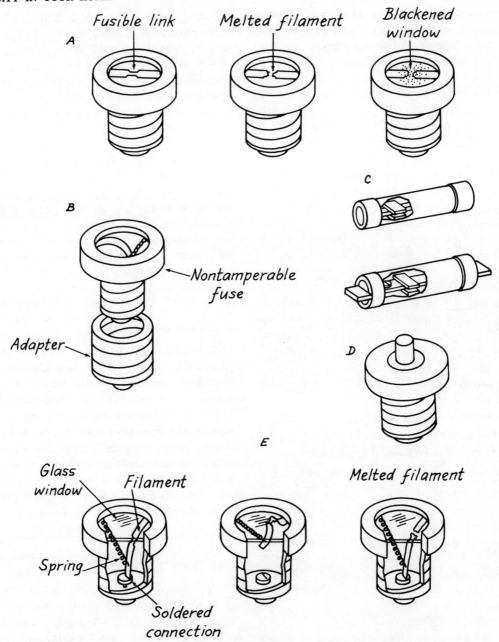

Fig. 13. The types of fuses found in home electrical systems: *A*. Plug fuses; *B*. Type S nontamperable fuses; *C*. Cartridge fuses; *D*. Resettable fuses; *E*. Time-delay fuses.

strip inside, but you can assume it has broken because of a short circuit.

TIME-DELAY FUSES

Outwardly, *time-delay* fuses look exactly like plug fuses. That is, they are round, threaded housings with mica windows in their tops. When you look inside them, however, there is a stretched spring attached to the fusible link, which in turn is soldered to the bottom of the unit.

Time-delay fuses were developed for use on circuits that have to bear a small overload for short periods of time (a few seconds). Suppose, for example, you have a 15-amp circuit protected by a time-delay fuse. There you are with perhaps 5 amps flowing through the circuit to light three or four lamps. Then you turn on a bench saw which is also supplied by the circuit. The saw motor draws about 30 amperes for the first few seconds while the motor gets up to speed, and then drops down to a steady demand of 6 or 7 amps. If the circuit were fused with a standard plug fuse it would blow the moment you turned on the saw. With a time-delay fuse, as much as half a minute will go by before the solder holding the fusible strip melts and the spring yanks the strip out of its position, breaking contact. During those first few seconds the circuit will be overloaded but the wire will not have time enough to overheat and cause a fire.

It is a good idea to install time-delay fuses on all circuits that supply large motor-operated appliances such as stationary power tools, air conditioners, and washing machines. If you examine a blown time-delay fuse and its window is clear, it blew because of an overload on the circuit. If the glass is smudged, there was a short circuit.

TYPE-S NONTAMPERABLE FUSES

When fuses begin blowing with some regularity, people have a tendency to replace them with higher-amperage fuses (or, even more dangerously, a copper penny) instead of reducing the load on the circuit. Homeowners play with the safety of their families this way all the time, but apartment dwellers are even bigger offenders (what do they care, they don't own the building, only their lives).

The *Type-S nontamperable* fuse is designed to protect people against themselves, and it comes in two parts. One portion is the fuse itself, which looks exactly like a time-delay fuse (with a spring-loaded fusible strip) but also has a metal spring at the top of its threaded portion, directly under the head of the unit. The second part is an adapter which is a threaded barrel that is screwed into the fuse holder in the distribution panel. Once inserted, the adapter cannot be removed without damaging the socket. Moreover, only the properly rated Type-S fuse can be used

in any given adapter. Thus, if the socket is designed to accept 15-amp fuses, the only fuse you can put in it is a Type-S 15-amp fuse or smaller. If the socket is designed for 20 amps, only the 20-amp and 15-amp Type-S fuse fits in it. Thirty-amp adapters will accept only 20-, 25-, or 30-amp Type-S fuses. In other words, when the adapter has been installed it is next to impossible to overfuse the circuit.

When inserting a Type-S fuse, rotate it into its adapter until it is as tight as you can get it. Then turn it some more, if you can. The spring on the underside of the fuse head must be flattened or the fuse will not make contact in the adapter. If you install a Type-S fuse and nothing happens to the circuit, you probably have not gotten the spring flat enough and you should tighten the fuse some more.

RESETTABLE FUSES

Resettable fuses look like standard plug fuses except they do not have a mica window. Instead, they have a button in their face which is connected to a switch inside the fuse. When the circuit that a resettable fuse is protecting is overloaded, the internal switch opens, breaking the circuit and pushing the button outward. All you need to do is push the button in to reset the fuse. Resettable fuses are sold with the same 15-, 20-, 25-, and 30-amp ratings as plug fuses. They are more expensive than other fuses, but theoretically they never need to be replaced.

CARTRIDGE FUSES

Any circuit with a rating over 30 amperes must have a *cartridge* fuse, although they are made in all ratings so you can install them on circuits under 30 amps. There are two basic types of cartridge fuses. The *ferrule-contact* type is a paper tube with copper caps on its ends and is manufactured in ratings up to 60 amps. The *knife-blade* types are also tubes but have tabs on both their ends and are used only in circuits carrying 60 amperes or more. Both types contain fusible links that melt when the circuit is overloaded and are held in place by spring clips that grip their metal ends.

Some versions of cartridge fuses have replaceable fusible links so they can be reused; both are made in standard and time-delay versions. However, the one-time fuses do not overheat their panelboard clips as much as the renewable ones and are generally considered more satisfactory. The one problem with cartridge fuses is determining whether a given unit has blown or not. There is no way of looking at the fuse to determine whether its fusible link is still intact, so it must be tested either with a neon hot-line test lamp, a voltohm meter, or a flashlight.

TESTING CARTRIDGE FUSES WITH A HOT-LINE TESTER

Use extreme caution when testing a cartridge fuse in a panel box. You cannot turn off the power in the box, so be very precise about where you engage the probes of the tester.

1. Remove the face of the panel box by undoing the screw or screws that hold it in position.

2. Touch one probe of the hot-line tester to the white return wire from the circuit cable and the other probe to the black wire on the incoming side of the fuse. If the tester light goes on, power is entering the fuse.

3. Keep one probe on the white wire and touch the other probe to the black wire on the exit side of the fuse. If the tester light goes on, power is exiting the fuse and it is functioning. If the light does not go on, the fuse has blown and must be replaced.

TESTING CARTRIDGE FUSES WITH A VOLTOHM METER

This is a safer procedure, but you have to have a voltohm meter to do it.

1. Remove the fuse from its clips.

2. Set the voltohm meter to the RX1 scale.

3. Touch the meter probes to the ends of the fuse. If the meter reads zero ohms, the fuse is good. Any other reading indicates that the fuse has blown and should be replaced.

TESTING CARTRIDGE FUSES WITH A FLASHLIGHT

If you have neither a hot-line tester nor a voltohm meter handy, you can still test a cartridge fuse using a flashlight, providing the light works to begin with. In other words, the batteries must be relatively strong.

1. Remove the fuse from its clips.

2. Remove the cap of the flashlight so that you can hold the bulb in one hand and the batteries in the other.

3. Hold one end of the fuse against the button on the top of the batteries in the flashlight case.

4. Press the bottom of the flashlight bulb against the other end of the fuse. You have to push them together hard. If the light goes on, the fuse is in working order. If the light does not go on, replace the fuse.

Fig. 14. When testing a fuse with either a meter or a hot-line tester, be very careful to touch the probes exactly where you want them. In this case one probe is at the exit side of the fuse. The other probe is at the white wire bus bar.

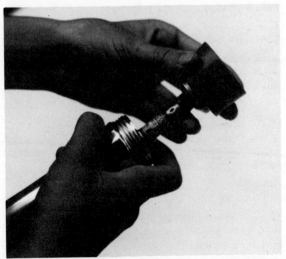

Fig. 15. How to test a cartridge fuse using a flashlight.

Cartridge fuses are often positioned in their clips on the back of a plastic cartridge. When you look at the panel board, the cartridges are discernible by the fact that they look like tiny rectangular doors with handles on them. The

Fig. 16. The cartridge fuses are held in clips behind the doors having handles. To get at the fuses, merely pull the handles toward you.

doors are usually stamped "Main" or "Range" or "Water heater." To get at the fuses, pull the handle straight out. The door of the cavity will come out of the panel, and when you turn it over you will find the fuses in their clips. It should be noted that if this is the case you cannot test the fuses with a hot-line tester, since they are no longer connected to any wiring system. You can, however, test them with a voltohm meter or a flashlight.

- To replace the fuse, grip it at its center and lift straight out, pulling it free of the clips that hold it at each end. Push the new fuse into the clips and then shove the fuse cover back into its cavity. When you are replacing the cover, be certain that you put it in right side up. The cover is

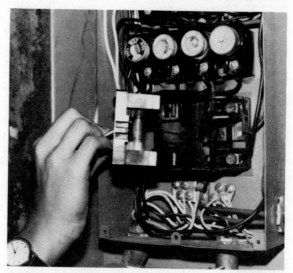

Fig. 17. When you have removed the cartridge fuses from the panel they can then be taken out of their clips.

marked "on" and "off" and so is the rim of the cavity. The "ons" should be next to each other.

CIRCUIT BREAKERS

A *circuit breaker* is a device that looks like a standard toggle switch, and in fact acts like any wall switch in your house. The lever on its face has "on" and "off" positions marked right next to it. When the lever is pushed to "on," the breaker is allowing current to flow through it; when the lever is in the "off" position, the breaker has "blown" and shut off current to the branch circuit. To reconnect the breaker, you simply push the lever to its "on" position (with some types, the lever snaps to a center or "tripped" position when it blows, and you have to push it to the "off" position first, then back to "on").

Circuit breakers rarely need to be replaced and offer a convenience (and safety) that has made them increasingly popular in recent years. Nearly all modern home constructions use them instead of fuses. Most types will carry 150 percent of their stated ampere rating for nearly a minute without tripping, and 300 percent of their stated load for about 5 seconds, which is long enough to carry the added current needed to start a motor. They are available in all of the same ratings as fuses.

Fig. 18. Circuit breakers look and act like any toggle switch. This particular breaker box has an open position; the fourth slot down from the top on the left side has no breaker in it and could be used for an additional branch line.

When used to protect 240-volt circuits, two breakers linked by a common handle are installed in the distribution panel. You will find these double breakers connected to the main service entrance, electric water heaters, ranges, and electric clothes dryers.

GROUND-FAULT CIRCUIT INTERRUPTERS

Often referred to as GFCIs or GFIs, *ground-fault circuit interrupters* are installed to protect either a complete 120-volt 2-wire circuit or individual receptacles against ground fault currents, which in effect are short circuits. There are various types of GFCIs, including ones that are plugged into a standard wall receptacle and others that are a combination receptacle and GFCI in a weatherproof casing for use outdoors. There are also types designed to be installed in a distribution panel and function as combination circuit breakers and GFCIs. The advantage of all

GFCIs, and the reason for using them, is that they will break the circuit they protect when as little as 5 milliamperes of current leaks into the ground by any path other than the return wire in the circuit. Not only do they react to an infinitesimal amperage, but they do it in about a fortieth of a second, which is quick enough to reduce what might be a fatal shock to a mild jolt.

GFCIs operate on a basic principle: If you have a normal 120-volt, 2-wire house circuit that includes a grounding wire, the same amount of current is flowing through the grounded wire as the hot line. But if the insulation in an appliance connected to the circuit is worn, or a wire comes loose from its terminal and touches the appliance housing, part of the current will flow through the bare grounding conductor, and some of it will go through anyone who is touching the appliance. Consequently, the amount of current in the return (white) wire will be less than it is in the hot line. A GFCI installed in the circuit will immediately react to the difference and open the entire circuit (if it is positioned to control the whole circuit) or open the hot wire to the receptacle (if it is installed to protect just the receptacle).

Installing GFCIs on every circuit or receptacle in your home is a good idea, and they are practically mandatory in any part of the house that is likely to be subjected to moisture, such as the bathroom, kitchen, laundry room and any outdoor circuitry. The National Electric Code and many local building codes specify that GFCIs be installed to protect all receptacles in bathrooms, swimming pools, and all 15-amp and 20-amp outdoor receptacles, for obvious reasons.

BRANCH CIRCUITS

Each of the fuses or circuit breakers in the distribution panel represents the beginning and end of a branch circuit that serves some part of your house. Physically, the circuits consist of a cable containing either two or three conductors (wires) plus an uninsulated grounding wire which leads from receptacle to receptacle (i.e., plug-in socket) and/or light fixture. The circuit, no matter what its load, must present a continuity of unbroken wires leading from the fuse or breaker to the farthest fixture or receptacle attached to the cable. Current flows from the main

Fig. 19. The power-feeder line enters this two-fuse box from the meter directly above the box. The branch cables come into the box from the bottom.

service cable through the fuse or circuit breaker and into the hot line (usually covered by black or red insulation) and is delivered to every receptacle, switch, fixture, and appliance on the circuit, then comes back to the panel box through the return, or neutral wire (always covered with white insulation). The current will go anywhere on the circuit that offers it a continuous path to travel. If it encounters an open switch (one that is turned to its "off" position) that controls a light or an appliance, it does not enter the device since its path is broken and not continuous. If the switch on an appliance is closed (turned to its "on" position), current will immediately flow into the unit, follow its entire circuit, and then return to the circuit cable and continue on its way. While all circuits in a house function in the same manner, there are three different types of branch circuits found in homes which are distinguished by their voltage, amperage, and purpose.

GENERAL-PURPOSE CIRCUITS

Homes constructed today usually have ordinary lighting circuits wired with #12 gauge wire and protected by 15- or 20-amp fuses (or breakers). Older homes typically use #14 gauge wire with 15-amp fuses. The general-purpose circuit feeds permanent lighting fixtures and most of the receptacles throughout your house. Generally, the circuits are laid out so that each one meets the electrical needs of about 400–600 square feet of living space. Another way of computing the length of a circuit is to reckon 3 watts (volts×amps) for every square foot of living space. Thus, if your home has 2,000 square feet of living area, and each 15-amp 120-volt circuit can provide 1800 watts, divide the number of watts by 3 square feet, giving you 600 square feet in each circuit, so your 2,000-square-foot home will need 3.3 circuits. But since you cannot have three tenths of a circuit, you will end up with four general-purpose circuits. That is a minimum requirement for electrical safety. In actuality, you would probably find that 5 or 6 circuits can better serve the day-to-day needs of an average family of four living in the house.

In any case, general-purpose circuits are meant to handle the relatively low electrical demand of the lights and small appliances found in most rooms.

SMALL-APPLIANCE CIRCUITS

Small-appliance circuits are normally cables with #12 wire, so they are capable of delivering 2400 watts and are rated for 20 amps. Their purpose is to supply power to areas of the house where various appliances are used, and they should not have any lighting attached to the circuit. Thus, small appliance circuits have only receptacles on them and there should be at least two such circuits serving any kitchen. Given a capability of delivering 2400 watts through each of the two circuits in a kitchen, you could safely use four appliances simultaneously. Pantries, dining rooms, breakfast rooms, laundries, and family rooms should also be supplied with at least one small-appliance circuit to be used strictly for small portable appliances such as heaters, fans, radios, etc. No stationary or fixed appliance, or lighting, should be plugged into them.

In keeping with the idea of safety, the circuits, or at least all receptacles in both the kitchen and laundry room, should be protected by ground-fault circuit interrupters.

MAJOR-APPLIANCE CIRCUITS

High-wattage appliances such as dishwashers, ranges, hot-water heaters, clothes dryers, and garbage disposals, any permanently connected appliance rated at 1000 watts or more (such as a bathroom heater), and any permanently connected motor rated at a ½-horsepower or more, should each be supplied by its own separate circuit. Some local electrical codes demand that the oil burner and blower be on separate circuits, as well as air conditioning units and solar heating systems. The circuits may be 120 or 240 volts, depending on the voltage of the apparatus they are feeding, and the wire must have enough ampacity for the current rating of the load it serves. The ampere rating of the overcurrent device depends on the appliance and its rating. If the appliance is rated at 10 amps or less, a 15-amp fuse or breaker is sufficient, although the current interrupter may be as much as 150 percent of the amperage rating of the appliance.

HOUSEHOLD ELECTRICAL NEEDS

Homeowners often assume that their houses are adequately wired to evenly distribute enough electricity throughout the house to meet all the needs of a modern household. Again, your ideal home has a 3-wire 120/240-volt system with 150-amp service, which by present-day standards will allow you to do anything electrically except heat or centrally air-condition the building (to include those capabilities, the service should be 200 amperes). But those are the modern ideals. Most homes have something considerably less than that. Many older homes have 2-wire, 30-amp service and perhaps even more are saddled with 60-amp, 3-wire service.

The chances are that since World War II the electrical demands of your family have increased steadily with the plethora of electrical appliances both large and small that have appeared on the consumer market. Nevertheless, if the needs of the household are less than 20 percent over the maximum capacity of the electrical system, you can still get along with what you have. If the maximum electrical needs are more than 20 percent of your present capacity, you may want to seriously consider upgrading and expanding your electrical service.

The maximum capacity of your house wiring system is usually (but not always) marked on the door of the distribution (fuse) box. The real capacity of the system can be determined by looking at the ratings on the main fuses or circuit breakers. When there is more than one main fuse or circuit breaker, add up their amperage ratings.

THE WIRING LIST

Before you decide to upgrade your electrical system in order to increase the building's electrical capabilities, you have to know your present electrical capacity. You also need to determine just how overloaded each branch circuit is, and estimate what you think the future needs of the household will be. The most lucid way of making this analysis is to work out a wiring list, which can not only delineate your present wiring and where each branch circuit goes, but will also tell you which appliances are on each circuit and how much current they draw. Making such a list is not difficult, just a little tedious.

HOW TO MAKE A WIRING LIST

1. Turn on all the lights in the house.
2. The fuses or circuit breakers in the distribution panel are usually numbered. Remove or shut off the current interrupter controlling the Number 1 branch line.
3. Go through the entire house and make a list of the location and wattage of any lights that are not burning. Test every receptacle by plugging a nightlight or neon test lamp into it. Also note which appliances are normally used on each receptacle and write down their wattage.

The wattage of light bulbs is stated on the bulb itself. Appliances may list their wattage on an information plate somewhere on the unit. If the wattage is not listed, the amperage will be, and you can find the watts by multiplying amps times volts (in most instances 120).

4. When you have completed your list for the Number 1 circuit, replace the fuse or reset the circuit breaker and turn off the power to the Number 2 circuit. Then go through the house again, noting all of the lights that are not burning, testing each receptacle, and noting the wattages on all appliances that are normally plugged into the circuit.

5. Continue testing each branch circuit until you have notated the entire electrical system and all of the lights and appliances in your house. As you make out the list, be sure to check such out-of-the-way places as the garage, the attic, and closets.

6. When you have completed your list, add up all the wattages written down for each circuit. Multiply the amperage of the circuit times its volts (120 or 240) and compare the wattage of the circuit with the total wattage of the appliances and lights used by that circuit.

USING YOUR WIRING LIST

The wattage of a 15-amp 120-volt circuit is 1800 watts; a 20-amp 120-volt circuit has a capacity of 2400 watts; 30-amp, 120-volt lines can deliver 3600 watts. It is unlikely that all of the appliances and lights on any given circuit will ever be running at the same time, but for the sake of safety the total potential power demand should not exceed 80 percent of the circuit's capacity. If you discover branch circuits in your wiring list that have lights and appliances connected to it that could potentially exceed the line's total potential wattage, try to move some of the appliances to another, less-loaded circuit. If that doesn't work, consider asking your utility to bring more power to your electrical system, or simply adding branch circuits.

When you look at the wiring list, particularly if you intend to add more circuits to your existing house system, bear future needs in mind. If, for example, you plan to turn the attic into bedrooms, or convert the garage into a family room sometime in the future, it is easier to run a branch line up to the attic or out to the garage while you are doing all of the rest of the work. If you are redoing the kitchen and plan to include a dishwasher or garbage disposal at some time in the future, it is easier to bring a major appliance line or two into the kitchen while it is under renovation than it is later when the room has been completed and decorated.

CHAPTER THREE

Tools and Equipment

There is only one rule of safety that anyone needs to observe when working with electricity: MAKE ABSOLUTELY CERTAIN THE POWER HAS BEEN TURNED OFF BEFORE YOU TOUCH ANY ELECTRICAL EQUIPMENT. So long as there is no current passing through any of the components you touch in a given circuit, there is no danger of electrical shock. Disconnecting the power may seem self-evident, but it is surprising how lazy people can be about going downstairs and unscrewing a fuse before they stick the metal blade of a screwdriver into a nest of 120-volt or 240-volt wires. Even at that, the fire insurance companies, electrical tool and equipment manufacturers, as well as the State, Local, and Federal governments, have done just about everything anyone can do to protect you from yourself. For example, most tools used for electrical work are sold with insulating rubber or plastic handles, and the electrical equipment installed in home wiring systems must all be listed by the Underwriters Laboratories. Even more importantly, the National Fire Protection Association has produced a series of standards and safety protection codes, among them the National Electrical Code (NEC), which is used as a safety guide by practically every electrician, and has in fact been enacted as a law in many communities.

Turning off the electricity before you begin tampering with a house branch circuit is a paramount rule, but safety really begins during the installation of electrical equipment. This involves properly grounding the separate components (preferably all) of your home electrical system. It entails installing ground-fault circuit inter-

rupters wherever the conductors have any chance of coming into contact with moisture. It includes using double-insulated tools and appliances whenever possible. And it also involves not being downright stupid about electricity.

Being smart about your electricity means remembering, at all times, that electricity can be dangerous to your health. Don't, for example, sit in your bathtub and blithely spin the dial of your radio, or turn on a hairblower or a heater; a short circuit in the appliance leaping into your body will become magnified to lethal proportions by the water. Don't even plug in a kitchen appliance when your hands are wet, or when you are standing in a puddle of water, or touching a water pipe or metal sink.

THE NATIONAL ELECTRICAL CODE (NEC)

The first year the National Electrical Code was published was 1897. At the time the Code was sponsored by the National Conference of Electrical Rules and was a list of rules and regulations for the safe installation of electrical materials. Publication of the NEC was taken over in 1911 by the National Fire Protection Association, and every three years or so since that time the National Electrical Code has been updated and revised.

The NEC represents the thinking and consideration of representatives from all parts of the electrical industry, and the committees that comprise the Code-making panels include recognized experts in all of the electrical fields. The work they perform has created an ongoing document often referred to simply as "the Code," which by its own statement is not intended as a specification or an instruction manual for "untrained" persons. The NEC intends to, and surely does, set forth a series of rules and provisions to make any electrical installation as safe as it can be made. The NEC has been adopted by the American National Standards Institute (ANSI), and the U. S. Occupational Safety and Health Act has, in effect, made the NEC a national law applicable to the types of businesses and industrial complexes that are covered by the Act. Moreover, a great many communities throughout the nation have adopted the NEC in toto or used it as a basis for their own electrical codes and standards. Once a city, county, or state governmental body legally adopts the Code, it of course becomes law, and compliance with it in that particular locality becomes mandatory by anyone doing any electrical work in their home, place of business, or anywhere else.

Without question, the NEC is a difficult piece of reading. It is written for professionals in the electrical industry as a safety code only, and makes very clear that it is not the intention of the NEC committee that the Code be anything but a safety code. In fact, the Code states that its suggestions may not even be efficient, or easy, or even adequate for good electrical service or the expansion of an electrical system to meet future needs. Nevertheless, anyone who is engaged in installing electrical wiring and equipment ought to possess the latest copy of the NEC. If the wording in it is sometimes confusing or apparently unclear, there are numerous unofficial guides and interpretations of the Code available at many bookstores. While these may or may not provide absolutely accurate interpretations of what the Code is saying, they can be invaluable aids to understanding what the NEC considers to be absolutely safe. To obtain a copy of the NEC, use the following address: National Fire Protection Association, 470 Atlantic Avenue, Boston, Mass. 02210. You can call 617-482-8755 for the current price.

LOCAL CODES AND STANDARDS

Practically every locality in the country has an electrical code or a set of laws governing electrical installation. The code may be a state law, or it may belong to a specific county, municipality, or town. Most of these codes and standards are based to one degree or another on the NEC, although they may be stricter (or more lenient) on certain points. No matter what they are, if you do any electrical work in your home, you are working under the auspices of that local electrical code, which, like the NEC, is principally concerned with the safety of any electrical system anyone constructs.

PERMITS AND LICENSES

In many locales, it is necessary to obtain a permit from the city, county, or state authorities before you do any wiring. There is normally a small fee attached to the permit, and most certainly any work that you or a professional electrician does will be inspected to be sure that your work meets all of the safety and installation standards of the local electrical codes.

In many places, the local law states that all electrical work must be performed only by a licensed electrician. In other areas, you are permitted to do much or all of the work yourself. But the only way you can know what your local laws permit is to ask your buildings department for a copy of the local code. For example, you may be allowed to install cables, receptacles, switches, and fixtures, but any work involving the distribution panel must be performed by a licensed electrician and be inspected by the municipality.

TESTING LABORATORIES

A municipal or county inspector examining your electrical work has neither the time nor the equipment to test each piece of equipment you have used in assembling your system, so he bases his approval of the components that make up your system on a listing of materials and equipment that have been approved by recognized testing laboratories. There are several such laboratories, including the Canadian Standards Association, the Underwriters Laboratories of Can-

ECTRICAL TOOLS

When you get down to doing the actual work,
can complete most electrical chores with a
of long-nosed pliers and a standard-blade
screwdriver. There are, however, a number of
other inexpensive tools that are helpful to have
available when you are working with your home
electrical system:

SAWS

Compass saws are ideally suited for cutting
holes in plaster, lath, or drywall when you want
to install an electrical box.

Hacksaws are absolutely necessary if you are
working with armored cable, which has a metal
sheathing.

cept products from any manufacturer, running
them through the most severe tests it can devise.
If they pass all of these tests, the items are listed
under one of various categories. Note that the
UL *lists* the products it has tested and found to
meet minimum safety standards. It does not approve any product. It only *lists* them, so it is
wrong to say an item is UL-approved, or anything that suggests that it is approved. By testing
and listing a given product, the UL has merely
been concerned with its safety, not with convenience or efficiency. And, in keeping with its very
strict considerations of safety only, the UL does
not list any product for use that would in any
way violate any of the safety requirements of the
NEC.

Safety, then, is everyone's concern in the electrical industry. The NEC sets forth rules and
standards for it. The UL lists only equipment
and materials that can be used safely (providing
they are used properly). Local and state electrical codes are designed for the safe installation of
all electrical equipment and materials. The inspectors who uphold those local codes and
standards follow the safety rule books. They will
reject any piece of equipment or any material
that does not appear on their copies of the UL
lists (which means they better all have a UL
sticker on them). They will not approve any installation that does not meet all of the standards
of safety as stated in their local electrical codes.

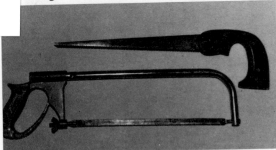

Fig. 21. A compass saw (top) and a hacksaw (bottom).

BORERS

Many electricians still use an old-fashioned
hand *brace and bit* when drilling through framing members so that cable can be fished through
the walls and ceiling of a house. Normally, the
practicing electrician will need an 18″ extension

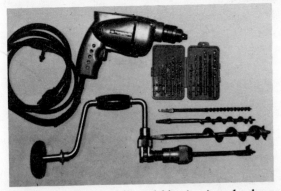

Fig. 22. An electric drill and bits (top) and a brace
and bits (bottom).

for the bits. However, most drilling chores can be done more efficiently with a hand power drill and some spade bits that can make holes large enough to accommodate electrical cable.

UTILITY KNIFE

A sharp *utility knife* with replaceable blades can be invaluable for all sorts of electrical and related chores. It can cut through drywall, be used to strip insulation from wires, and cut non-metallic-sheathed cable.

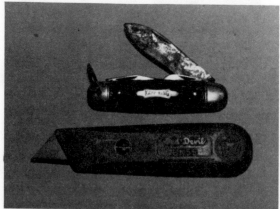

Fig. 23. A jackknife (top) and a utility knife (bottom), which has replaceable blades.

STRIPPERS

There are various types of *wire strippers* on the market, and any of them will do the job of neatly stripping insulation from the ends of various gauge wires. You don't absolutely need a wire stripper, but it will more than pay for itself in time saved, to say nothing of the fact that strippers do not knick the copper wire (and therefore weaken it) the way a knife will.

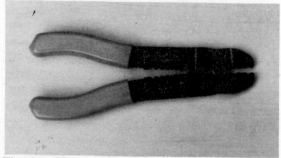

Fig. 24. The multipurpose tool will cut and strip wire, cut bolts, crimp insulation, and crimp-on components.

A *multipurpose tool* is even handier to have than wire strippers because it not only strips insulation but also can be used to cut wire as well as crimp terminals and fasteners on the end of a conductor without the use of solder.

PLIERS

There are all kinds of pliers available. The most versatile of them, so far as electrical work is concerned, is probably the long-nosed (or alligator) pliers, since their jaws are narrow enough to bend wire into the small loop necessary to fit around terminal screws, as well as grip small screws and other electrical parts. Most long-nosed pliers also have cutters along the sides of their jaws that can be used for snipping off wire. A standard pair of *slip-joint pliers* is a good backup to have in your tool kit, and if you are really doing a lot of electrical work, a pair of *electrician's pliers* can not only cut wire but will strip it as well. *Channel-type pliers* come in different sizes and provide considerable leverage, as well as the ability to get at hard-to-reach nuts and bolts. *Locking* or *vise-grip* pliers have the distinct advantage of being able to be locked on a nut and hold it in place, freeing you to use both hands to do something else.

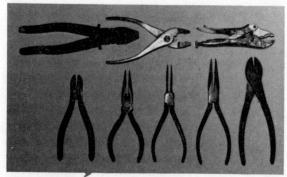

Fig. 25. There are all kinds of pliers and cutters that you can purchase and they will all do the basic job of bending wire and holding things.

WRENCHES

A standard, medium-sized *adjustable-end wrench* can hold any number of large and small nuts and often becomes the only tool that will free a frozen nut. A set of small combination *box and open-end wrenches* is also useful, especially if you are dealing with appliances and motors.

With appliances you may encounter *allen screws,* which can only be tightened or loosened with *allen wrenches.* You can buy allen wrenches singly or in sets that include several diameters. *Nut drivers* come in sets containing both large and small nut sizes. For electrical work primarily involving appliances, a set of small drivers can be useful for tightening and loosening out-of-the-way nuts.

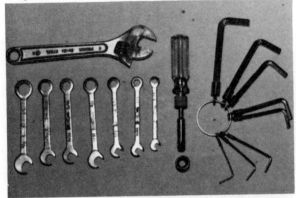

Fig. 26. Wrenches can be invaluable; your electrician's kit should include an open-end wrench (top), a set of small open- and closed-ended wrenches, a nut turner (center), and a set of allen wrenches.

SCREWDRIVERS

The simplest way of meeting your screwdriver needs is to buy a set of seven screwdrivers. The set will include four different standard-blade

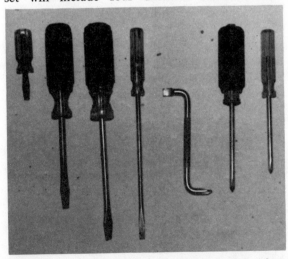

Fig. 27. A set of seven screwdrivers includes four standard blade tools with different-sized blades, two phillipshead screwdrivers, and a small S-shaped tool that has a standard blade on one end and a phillipshead on the other for use in tight situations.

screwdrivers, one of which should handle just about any screw you encounter in electrical work. There will also be two different-sized phillipshead screwdrivers, which have blades in the shape of an X. Phillipshead screws are used by many appliance manufacturers but not in home wiring equipment. The set should also include a small S-shaped screwdriver with a standard head at one end and a phillipshead at the other, which can be used to get at hard-to-reach screws. Buy a set that has rubber handles; they are easier to grip and considerably safer if you happen to touch the metal blade to a hot wire.

NEON TEST LAMP

This costs about one dollar and contains a small neon bulb that glows when the probes attached to it are touched to current-carrying parts of an electrical system. You can use them to check wall outlets, for example, but be sure you buy one that is rated for 120/240 volts. Some units are manufactured for lower voltages and will burn out if you use them to test your house electrical system.

ELECTRICAL EQUIPMENT

Unlike plumbing or woodworking, electrical work around the home does not require a great many different types of equipment or materials. All splices, switches, and receptacles must be housed in a metal or plastic box, and all of the boxes must be interconnected with electrical

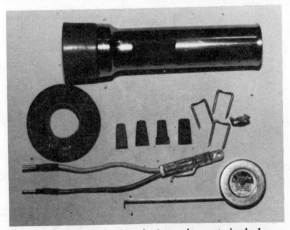

Fig. 28. Assorted electrical equipment includes a flashlight and electrician's tape, wire nuts, U-nails, a hot-line tester, and a tape measure.

cable. Aside from a very few other small pieces of equipment to effect these connections, there is no other equipment to be considered. Remember that every component you install in your house must be UL-listed or your wiring system will not meet the minimum safety standards as described in the NEC.

CABLE

Electrical cable is classified according to the number of wires it contains and their gauge size (Fig. 29). Two-wire cable is, in actuality, made up of three wires. The first wire is covered with white or grayish insulation; this conductor is always to be used as the return, or neutral, wire. There is a second wire wrapped in black insulation, signifying that it is always to be used as the "hot" line. The third wire is usually uninsulated and is to be used as the grounding wire.

Three-wire cable consists of three wires plus a bare grounding wire. One wire is white, one black, and the third may be black or red (or any other color except green). The black or red third wire is also a hot line.

The sheathing around the wires can be nonmetallic or armored, and all cables are clearly labeled with letters and numbers that define what the cable is. The numbering system gives the wire gauge (for example, #14 or #12) and the

number of wires in the cable (for example, 2 or 3). Thus, a cable might be labeled #12/2, meaning the cable consists of two #12 wires. The maximum voltage the cable can handle is also stamped on the sheath, along with whether the product is UL-listed. A typical cable stamping appears every two or three inches along its sheathing, and might read "#12/2 (UL) Type NM with ground 600 volt." In other words, the cable is type NM (nonmetallic) and contains two #12 wires with a bare grounding wire, and it is capable of carrying 600 volts. It should be noted that although it is *capable* of carrying that many volts, the NEC does not permit it to bear anywhere that much of an electrical load.

NONMETALLIC-SHEATHED CABLE

Nonmetallic-sheathed cable essentially contains electrical conductors (i.e., wires) that are sheathed in rubber or plastic. Nearly all of the cables made today use plastic. There are several types of nonmetallic-sheathed cables acceptable for use in home wiring systems in specific circumstances.

Type NM is the most common of all cables and may be used in any permanently dry location. Each wire is insulated with plastic and then wrapped in a spiral of paper for extra protection. The sheath itself is moisture- and flame-resistant,

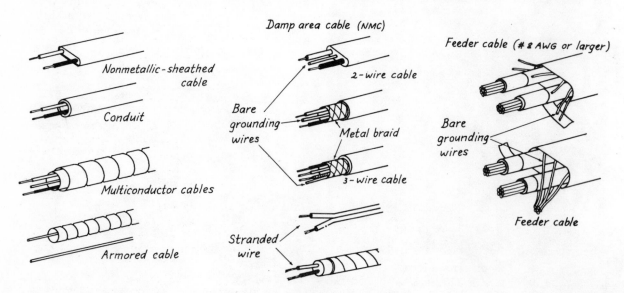

Fig. 29. Different types of cable.

and can be either a fibrous or a plastic material. NM can be used all over your house except in damp or corrosive locations.

Type NMC is designed for use in dry, damp, wet, or corrosive areas. Individual wires in the cable are embedded in solid plastic, which makes them all waterproof, although NMC cannot be placed directly in the ground.

Type NMC is not readily available in every locale, but the more widely distributed Type UF (Underground Feeder) is considered an excellent alternative. It can be used wherever NMC is used and can also be buried in the ground, providing there is overcurrent protection at its starting point.

Perhaps the greatest practical asset to using nonmetallic-sheathed cable is its flexibility and the ease with which it can be cut, stripped, and clamped to various electrical boxes. However, not all locales permit its use, in which case your only alternative is armored cable.

ARMORED CABLE

Armored cable is usually referred to as "BX," but that is actually a trade name. The Code refers to it as Type AC or ACT, depending on the kind of insulation used to protect the individual wires. Each wire in an armored cable is covered by a plastic (ACT) or rubber (AC) insulator (white, black, red) and all armored cable is now sold with a bare grounding wire as well. The wires are wrapped in paper to protect them from any abrasion from the metal sheath, which takes the form of a spiral of metal. The armor must be cut with a hacksaw, and at best it is nowhere near as flexible as its nonmetallic-sheathed brethren. It does have the advantage of forming its own grounding system via the metal armor, however.

CONDUCTOR/WIRES

Conductors conduct electricity from its source of voltage to wherever it is used. The most common form that a conductor takes is wire, although it is sometimes a bus bar or some form of metal strip suitable for carrying current. All conductors have a certain amount of resistance to the flow of current that prevents the current

from flowing through them with total freedom. That resistance inevitably causes voltage drop. Consequently, for any given electrical load to receive the amount of current it needs to operate, the size of the wire must be taken into account, as well as how far the wire must extend from its source to the load.

The size, whatever it may be, must be large enough to limit voltage drops, but there is also the problem of heat. Current flowing through a wire causes heat, which varies according to the square of the current (amperes). Just how much heat various types of insulation can withstand, before it reaches a temperature high enough to cause a fire, is of grave concern to the NEC. The Code therefore states the ampacity (maximum current-carrying capacity in amperes) that it considers safe for every size of wire manufactured. The wire gauges normally used in residential wiring and their ampacities are:

WIRE GAUGE	AMPACITY
#14	15 amp
#12	20 amp
#10	30 amp
#8	40 amp
#6	55 amp

All of the wires found in residential wiring are capable of handling up to 600 volts (except the fixture wires used in the internal wiring of lights, which are capable of handling only as much as 300 volts). Most wires have a thermoplastic insulation around them of a thickness that depends on the diameter of the wire. This insulation is usually described on either the insulation or the

Type	Locations	Sizes	Temperature ratings
RHH	Dry only	Nos. 14, 12, 10	75°C or 167°F
		No. 8 and larger	90°C or 194°F
RHW	Dry or wet	All sizes	75°C or 167°F
T	Dry only	All sizes	60°C or 140°F
TW	Dry or wet	All sizes	60°C or 140°F
THW	Dry or wet	All sizes	75°C or 167°F
THWN	Dry or wet	All sizes	75°C or 167°F
THHN	Dry only	Nos. 14, 12, 10	75°C or 167°F
		No. 8 and larger	90°C or 194°F
XHHW	Dry only	Nos. 14, 12, 10	75°C or 167°F
	Dry only	No. 8 and larger	90°C or 194°F
	Wet only	All sizes	75°C or 167°F

Fig. 30. Temperature ratings of ordinary types of wire.

cable sheathing in terms of a letter or group of letters (T, THW, THHN, THWN, etc.) which designates which conditions the wire can be used in (damp, dry, wet). Type T can only be used in dry conditions. Type THW can be used in both dry and wet conditions. THHN is acceptable only in dry conditions, and THWN can be used in either dry or wet conditions.

When wire is discussed in most electrical wiring contexts, it is assumed to be copper, since that is what has long been in use, and still represents the majority of installations in homes today. The alternative to copper is aluminum, but aluminum has caused some problems that make it forbidden by many local electrical codes.

Aluminum has a higher resistance than copper, and therefore must be larger in diameter to carry the same number of amperes. The rule of thumb is to use an aluminum wire two sizes larger than copper to get the same ampacity. Thus, if you would normally use ⚡12 copper wire, go to ⚡10 aluminum.

The major problem with aluminum wire occurs when attaching the wire to connectors or terminals. Aluminum expands faster and considerably more than copper, so when you attach an aluminum wire to a copper or brass terminal screw, the screw is expanding and cooling at a slower rate than the aluminum wire, and eventually the connection will work loose. Not only is that annoying, but it has been known to cause fires. The only safe remedy is to use UL-listed devices that are labeled for specific use with aluminum wires and designated as CU-AL, AL, or CO-ALR. CU is the chemical abbreviation for copper and CO is an arbitrary marking that means the same thing. AL stands for aluminum, and ALR means aluminum revised. Any receptacle, switch, connector, or other electrical device with any of these stampings has been designed to be used with copper, copper-clad, and/or aluminum wire. If you don't use such devices, you are in danger of having the connections work loose and perhaps causing an electrical fire.

The dangers from aluminum wiring are real enough so that if your home has aluminum wiring you would be well advised to spend the time and money to change your electrical system. At the very least, replace any switches and outlets that are not properly rated with CU-AL, AL, or CO-ALR rated devices. If at all feasible, rewire the entire house.

WIRE GAUGES

The diameter of a wire is known as its gauge, and the gauges have been established by the American Wire Gauge (AWG). Wire gauges work backward: the larger the gauge number, the smaller the diameter of the wire. The numbers which are commonly used in house wiring are ⚡14 (the smallest), 12, 10, and 8. Numbers 16 and 18 are primarily found in flexible cords, signaling systems (intercoms, door buzzers), and other areas where small currents are needed. All of these sizes are made up of single-strand wire. From Number 6 up to Nos. 0, 00, 000, 0000, and larger, the wires become too large in diameter to be flexible enough for practical applications, so they are braided together from smaller wires. When you are working with your house wiring system, the gauges you are primarily concerned with are ⚡10, ⚡12, ⚡14, ⚡16, and ⚡18 copper wire.

WIRE COLORS

Wire is not colored. Wire is always the color of what it is—copper (or aluminum). The insulation around the wires is colored, however, and those colors indicate very specific purposes. White wires (with one exception concerning the connection of certain switches) are always used as the grounded return wire, and no other wire may be white. Black, red, blue, or yellow wires are used as hot wires. Bare (uninsulated), green, or green with yellow stripes are always used as the additional grounding conductor.

TERMINAL COLORS

Wires are connected to terminals, which are usually screws attached to various electrical devices. Natural copper or brass terminals are always used to hold hot wires. White-colored terminals that are nickel, tin, or zinc-plated are used to accept grounded (white) wires only. If there is a green-headed screw on a switch or receptacle, or any device, it is to be used only for the green or bared grounding wire.

CONDUIT

There are several types of conduit, but all of its forms amount to pipe made of steel or sometimes aluminum that has been specially manufactured for use in wet or corrosive locations. In other words, you use it outdoors. Once it has been assembled to various receptacles, switches, and fixtures, electrical wires are fished through it and are consequently protected from the weather.

Conduit can be cut with a hacksaw or a pipe cutter. By using a conduit bender, commonly known as a *hickey,* it can also be bent to almost any angle desired. Conduit is sold in 10' lengths, and the Code prohibits more than four 90° bends in any run between any two openings. The Code also prohibits any splices of any kind inside any conduit. The wires must be continuous through the conduit and breaks can only be made inside a switch, outlet, or junction box. The Code also regulates how many wires can be run through a conduit, depending on the size of the wires and the diameter of the conduit.

Fig. 31. Conduit is a pipe that contains electrical wire. An outdoor switch might have conduit leading into the weatherproof switch box and then out of it again.

ELECTRICAL BOXES

All wire splices and connections must be made inside an electrical box, to provide a maximum safety from both fire and shock. So, every time a cable sheathing ends, it enters an electrical box where its wires extend several inches beyond it

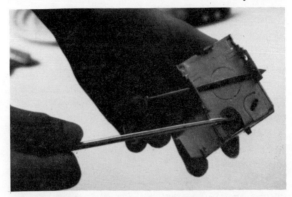

Fig. 32. The knockouts in an electrical box can easily be punched out by hitting them with a screwdriver. Some knockouts have a slot in them that makes prying them loose easier.

and either are spliced to other wires or end at the terminals on a switch or outlet.

Electrical boxes come in a variety of sizes and shapes, and all of them contain circular *knockouts* around their sides and bottoms. The knockouts are held in place by a single burr of metal, and are round disks that are easily pried out of the face of the box with a screwdriver. The hole is exactly large enough to hold a cable clamp that will rigidly secure any cable entering the box. The NEC lists the maximum number of cables entering each size of box, and is also very specific about how boxes must be attached to the framing members of your house so they will not shift under the weight of whatever equipment they contain. Moreover, if a knockout is removed from an electrical box and then not used, the Code prescribes that the hole be covered with a special knockout closure.

While the majority of residential installations still use metal boxes, nonmetal (plastic) units are now available and can be used with Types NM, NMC, and UF cables. One of the advantages of the plastic boxes is that cables entering them do not need to be clamped, providing the cables are securely anchored within 8" of the box.

The boxes you use in any given situation, whether they are metal or nonmetallic, depend on the number of wires that must enter them, and the purpose of the box.

OUTLET BOXES

Outlet boxes can be octagonal (4″ in diameter) or square (4″ and 4¹¹⁄₁₆″ in size) and can be either 2½″ or 2⅛″ deep (Fig. 33). While they are called outlet boxes, they can actually be used to hold receptacles, switches, or light fixtures, provided they have the right cover. A fixture or other device mounted on the face of a box that completely covers the box is considered a cover. Otherwise, a cover must be attached to the box opening. The covers available can be solid metal or plastic plates, or one that has a cutout in it to receive the face of one or two receptacles or switch levers. There are also dozens of different covers for octagonal boxes, making it easier to install ceiling fixtures of various types. Octagonal boxes are most often used in ceilings to support overhead light fixtures.

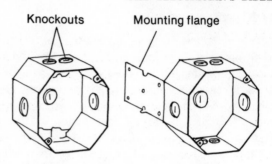

Fig. 33. Octagonal outlet boxes.

SWITCH BOXES

Switch boxes are rectangular and available in depths of either 1½″ or 3½″ (Fig. 34). Some switch boxes come with metal flanges attached to their sides to be used when installing the box next to a framing member (joist, stud). You can also remove the sides of a switch box so that two or more boxes can be hooked together. Ganging them like this permits several switches or other devices to be positioned in the wall next to each other.

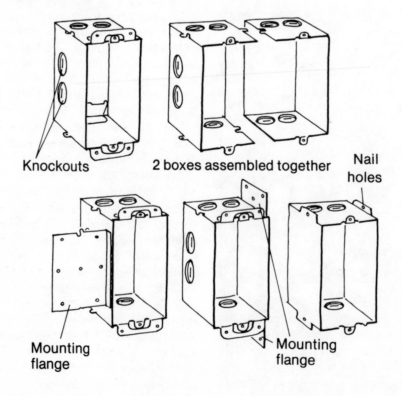

ELECTRICAL BOX ACCESSORIES

All switch boxes, as well as outlet boxes, have tapped holes which can be used to attach terminal screws for grounding wires, and to drive nails or screws through when attaching the box to house framing members. Some boxes are sold with single and double cable clamps already in the box, but there are also separate clamps that are inserted through any knockout in any box and are held in place with a locknut. The clamps are specially designed to hold either armored or nonmetallic-sheathed cable, so be sure to buy the type that is suitable for the kind of cable you are installing.

There are numerous types of adjustable hangers available to secure ceiling boxes between joists (crossbeams on the ceiling) so they will be strong enough to support an overhead fixture, and there is a complete assortment of fiber rings, metal clamps and adapters to be used with conduit to completely weatherproof any outdoor circuit (Fig. 35). All box covers, whether they are full covers or are used to protect outlets or switches, are also available in special weatherproofing designs to be used in outdoor installations.

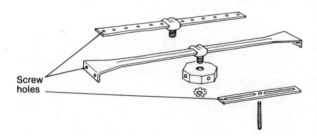

Screw holes

Fig. 35. Adjustable hangers are designed to span the tops of joists.

WIRE NUTS AND ELECTRICIAN'S TAPE

Wires, when they are connected to each other in a house system, cannot be soldered because an overheated wire might melt its connection. They must be spliced together and wrapped with electrician's tape or held with wire nuts. The tape you buy now has a plastic topping which is more heat resistant than the old electrician's tape.

Wire nuts come in two versions. The most common type is a plastic cone with a small copper wire coiled inside it that needs only to be screwed down over the wires after they are twisted together. The second type has no plastic covering over its spring, and must be insulated with electrician's tape once the wire connection has been made.

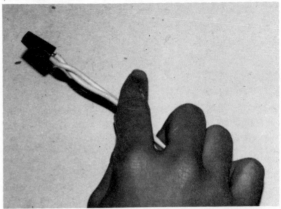

Fig. 36. Wire nuts are available in different sizes to accommodate the number of wires and their sizes to be spliced together.

RECEPTACLES

Lamps and portable appliances have to be plugged into something, and that something is a wall receptacle. The plug on the appliance cord has two prongs that fit into parallel slots in the face of the receptacle. The result is that one side of the appliance cord receives current through the black (hot) wire attached to the back of the receptacle in the electrical box. The other prong makes contact with the white wire, which is also connected to the back of the receptacle (Fig. 37).

Many older homes that were electrified before the 1960s have receptacles containing only two slots in their face. Now, all receptacles are made with a third hole centered between and below the slots to accept the grounding prong in many appliance plugs. In this manner, the appliance is grounded to the house grounding system and provides added safety to users from harmful shock. The receptacle is, in turn, wired to the

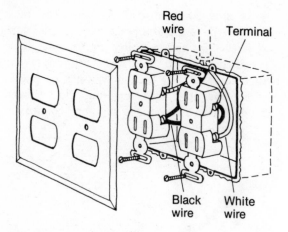

Fig. 37. How a quadruple electrical outlet is wired. The grounding wires have been omitted from the drawing for clarity.

cable entering the box that holds it, and is also grounded to the box with a green or bared grounding wire. Any time you have occasion to replace a receptacle, install a grounded outlet for it also.

You can purchase single receptacles that contain only one pair of slots, but most people these days double their options for running appliances by installing duplex receptacles, which consist of two sets of openings to accommodate two plugs at a time. These are called quadruple outlets.

There are also safety outlets. These have curved slots so that a plug must be rotated as it is inserted in the slots before it can seat properly. The safety outlets are almost childproof. Receptacles are also manufactured in combination with perhaps a switch, or a switch and a night-light socket, that will fit into any electrical box. In fact, since all switches, receptacles, and light fixtures are manufactured according to specific standards, they will all fit into any electrical box, new or old. All switches are rated according to their amperage and the number of volts they can handle without burning out. Be sure that you don't, for example, put a 15-amp, 120-volt unit on a 30-amp, 240-volt circuit, or it will be destroyed almost immediately.

SWITCHES

The purpose of any switch is to provide the user with a convenient way of disconnecting the electricity coming from its source to a light or appliance. Switches all have a lever or button which, when you move it, completes or breaks contact between the wires in the circuit. Electricity will not flow through the switch until it is closed (turned to its "on" position).

While the outward appearances of switches can vary considerably, they all function in the same manner. Here is a description of each of the basic switch types:

TOGGLE SWITCHES (SINGLE-POLE)

Toggle switches are completely enclosed units that have two terminals and the words ON and OFF on their handles. They are the most common type of switch found in homes. Both of their terminals are brass, and only the black wire of the cable that feeds them (and leads to what-

Fig. 38. A modern, 3-hole receptacle, or outlet.

Fig. 39. A standard single-pole toggle switch.

ever light or device the switch controls) is attached to the switch. The switch thus opens and closes only one wire in the circuit and is therefore known as a single-pole switch.

DOUBLE-POLE SWITCHES

These can look outwardly like a single-pole toggle switch, but in fact they have four terminals instead of two, and are usually referred to simply as double-pole single throw (DPST) types. They have the words ON and OFF on their face. Because it has four terminals, a double-pole switch is designed to open and close two wires in a circuit simultaneously with a single "throw" of the switch lever. The NEC requires that double-pole switches be used on 240-volt circuits, in which case each end of the two hot lines in the circuit are connected variously to the four terminals on the switch.

THREE-WAY SWITCHES

If you want to have the flexibility of turning on a light from either the top or the bottom of a stairway, a 3-way switch must be used at each control position. A 3-way switch can look like an ordinary toggle switch except that it has three terminals and the words ON and OFF are *not* printed on the face of the unit.

When you install 3-way switches, you actually have to use two of them in conjunction. They are wired so that the light they control will be on at any time the switch levers are pointed in the *same* direction. If, for example, both levers are pointing up (or down), the light will be on. But as soon as either lever is moved to an opposite direction, the light will go off. When wiring the switches, the hot line from the circuit is connected to the center terminal on one of the switches, and the hot line from the light is connected to the center terminal on the other switch. Separate black wires known as *travelers* are then used to connect the two top terminals on the switches, and the two bottom terminals. So long as the levers on the switches are going in the same direction, they allow the current to flow from the source through either the wire connecting the top terminals or the one connecting the bottom terminals. As soon as one of the levers is

Fig. 40. A careful look at the three switches shown here will reveal that the one on the far left has three terminal screws. It is the extra terminal that allows it to function as a 3-way switch.

pointed in the opposite direction, each switch is closing one of the connecting wires and also opening the other, so there is no longer any uninterrupted path for current to follow to the light.

FOUR-WAY SWITCHES

Now suppose you want to control a stairway light at the top of the stairs, at the bottom, and at the second-floor landing. At the point nearest the light and also at the point nearest the source of voltage, you must install 3-way switches; the middle position, however, gets a 4-way switch. The 4-way switch has four terminals and the words ON and OFF do not appear on its face. The switch is connected in the middle of the travelers connecting the two 3-way switches, and is designed either to allow the traveler wires to function as if the 4-way switch were not there at all, or to cross the wires so that the circuit will open and no longer carry current to the light.

There is no danger of shock or fire if, as you install a switch, you put the wrong wires on the wrong terminals. The circuit simply will not work. You need only proceed to try different wires on different terminals until the system is operating properly. *Just remember that no current must be in the circuit as you work on it.*

SPECIAL SWITCHES

There are numerous kinds of special switches available. Some are lock types that can only be operated by using a key that fits into the unit. Some types are known as momentary-contact switches which will only allow current through them for as long as the handle is physically held in its ON position. The moment you let go of the lever, a spring returns it to the OFF, or open, position. Some switches have buttons instead of a lever; others can be surface-mounted, to the face of a garage or basement wall.

SWITCH RATINGS

Switches are rated according to the maximum number of amperes and the maximum number of volts they can handle safely. A common rating, usually stamped on the face of the unit, might be 10A/⋇125V-5A/⋇250V, meaning that the unit is capable of handling as much as 10 amperes at no more than 125 volts, or up to 5 amperes at no more than 250 volts.

CHAPTER FOUR

Basic Working Procedures

Any time you are about to work with any electrical system, begin by turning off the power. Then check the circuit to make absolutely certain it does not have any electricity coming into it. DO NOT ASSUME YOU HAVE REMOVED THE SOURCE OF POWER. Test the circuit and be *absolutely* certain. There are a number of situations when simply removing a circuit current interrupter may not be enough. For example, if you are working on a ceiling fixture, the electrical box that holds the light may also be a junction box for more than one circuit; to nullify all of the power entering the box, you may have to disconnect the current interrupter to more than one circuit. Another situation, when just removing the circuit interrupter is not enough, occurs with the capacitor-start motors, commonly found in refrigerators and stationary power tools. The capacitor is a storer of electricity. You can unplug the motor but electricity still remains in the capacitor, so it must be separately deactivated (see Chapter Nine).

So always remove the current interrupter from its circuit, and always test the receptacle, switch, or fixture with a hot-line tester (neon test lamp) before you begin any work. If a receptacle is to be worked on, plug an appliance into it, or insert the probes of a hot-line tester in the slots of the outlet. If you are working on a wall switch, turn it on and off several times to be sure it no longer activates the fixture it is connected to. If you are entering a ceiling fixture, carefully remove the wire nuts covering the white and black wires. Then just as carefully touch the probes of a hot-line tester to the bared wires (one probe to the

Fig. 41. If you insert the probes of a hot-line tester in the slots of a receptacle and the tester light goes on, the receptacle is "hot."

white wires and one to the hot ones). If the neon light goes on, locate the other source of electricity before you touch any of the wires. In the interest of safety, get in the habit of using plastic- or rubber-handled tools, so that if you do touch a hot wire with their metal ends, electricity will not enter your body.

If you must work in a damp area, stay out of puddles and wear rubber-soled shoes. Standing in water when your body receives any kind of an electrical shock can intensify that shock to fatal proportions.

Any time you have occasion to remove the face of a distribution panel, first disconnect the main power interrupter, and still be very careful. If the main power disconnect is a part of the distribution board, even with the main current interrupter off, the feeder cable that comes from outside the house remains alive and very dangerous. Aside from being extremely careful about where you place the metal blade of a screwdriver or the jaws of your pliers, keep one hand in your pocket. That way you will not be tempted to touch anything metal, or damp, and you may save your life.

HOW TO CHANGE A FUSE

Fuses blow all the time in homes and apartment buildings all over the country. And all over the country people open the door to their distribution boxes, locate the blown fuse, and promptly replace it, without disconnecting the main circuit interrupter. That is dangerous, since you are sticking your fingers into a metal box that is full of electricity. Instead:

1. Disconnect the main current interrupters, shutting off all power entering the fuse box.

2. Locate the blown fuse and unscrew it by rotating it counterclockwise in its socket.

3. Shut off all lights and appliances connected to the circuit.

4. Thread a new fuse having the same amperage rating as the old fuse in the fuse socket by rotating it clockwise until it can no longer be hand-tightened.

5. Reactivate the main current interrupters.

6. Turn on the lights and appliances on the circuit one at a time, particularly if the fuse has blown because of an overload in the circuit. If the mica face of the blown fuse is blackened, there was a short circuit, most likely in one of the appliances. Be careful to turn on the appliances one at a time. If the new fuse blows and its mica face turns black when you activate a particular appliance, stop using the device until it has been repaired.

HOW TO RESET A CIRCUIT BREAKER

For absolute safety, you should also turn the main circuit breaker off before you reset a blown circuit breaker. However, circuit breakers and the panel boxes where they reside are designed with maximum safety in mind, so you can be relatively secure about just resetting the breaker. As with any time you touch a distribution panel, be certain that you are not standing in water or touching anything that might ground you in case you get a shock. Reset the circuit breaker as follows:

1. The handle of a blown circuit breaker will have moved away from its ON position. It may be in a center "reset" position, or have snapped to OFF. If the lever is in a center position, push it to OFF.

2. Turn off all lights and appliances connected to the circuit.

3. From the OFF position, push the lever to ON.

4. Turn on the lights and appliances connected to the circuit one at a time. If the circuit breaker goes off at any time during this procedure, particularly if it blows as you turn on an appliance, there may be a short in the circuit or appliance. If the breaker does not go off until you have reactivated all of the equipment attached to the circuit, you are probably overloading the circuit and should move some of the appliances to another circuit.

PROPER USE OF AN ADAPTER PLUG

Adapter plugs are not very expensive and can be purchased at almost any hardware store. People who live in old buildings buy these plugs all the time because old buildings have wall receptacles with only two slots in their faces, which is not enough holes to accept the 3-prong plugs attached to many modern tools and appliances.

Adapter plugs have two prongs on one end that will fit into any receptacle, and two slots and a round hole on the other, to accept 3-prong plugs. They also have a green pigtail (short wire) with a U-shaped connector crimped to its end. The purpose of the adapter plug is not so you can use your 3-prong appliances in a 2-holed receptacle. The purpose of the adapter plug is to protect you from electrical shock. But it cannot give you any protection whatsoever unless you *use* the green pigtail.

The pigtail is colored green because it is a

Fig. 42. In order to provide any safety, the adapter plug pigtail must be attached to the screw in the faceplate of the receptacle.

grounding wire, and it is connected to the third hole in the 3-hole end of the plug. The third hole, by the way, receives the third, round prong on 3-prong plugs, and that third prong represents the end of the grounding system inside the appliance or tool. Do *not* cut the prong off just because you need to use the appliance and have no adapter handy. If you cut it off, you have removed all of the safety margin that the grounding wire inside the equipment is designed to give you. To carry that grounding system from the appliance through the house wiring system and into the earth outside your house, the pigtail on the adapter *must* be connected to the screw in the faceplate of the receptacle, as follows:

1. Plug the adapter into a receptacle.

2. Loosen the screw in the center of the faceplate enough so that you can slide the U-shaped connector on the end of the pigtail under it. Then tighten the screw.

3. Plug the 3-prong plug into the adapter.

USING SCREW TERMINALS

Wire is usually connected to switches, receptacles, and other devices by means of a screw ter-

minal. The screws are inserted through a brass plate which is surrounded by insulating material. Most terminal screws have been flanged or "upset" at their bottoms so they cannot be entirely removed from their holes. Screw terminals are supposed to hold only one wire at a time. In fact, the NEC prohibits more than one wire under any single screw, since there is every chance the screw will not hold more than one wire securely. Nevertheless, there are numerous occasions, when wiring a house electrical system, when you will have the urge to ignore the NEC safety precaution and tuck two or three fat little wires around a skinny little standard terminal. Don't give in to the temptation. If you have two or more wires that must be connected to the same screw terminal, do this instead:

1. Cut and strip the insulation from both ends of a short length of wire. The wire should be the same gauge as the wires to be connected to the terminal.

Fig. 43. Strip insulation from both ends of a short length of wire.

2. Twist one end of the short wire together with the wires to be connected and hold them together with a wire nut.

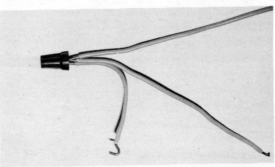

Fig. 44. Splice the short length of wire to the wires to be connected to the terminal.

3. Connect the free end of the short wire to the terminal.

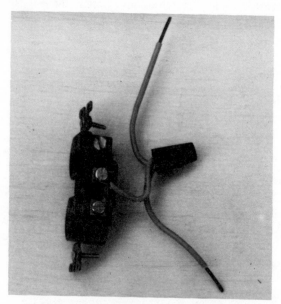

Fig. 45. Connect the free end of the short wire to the terminal.

PREPARING WIRE FOR A TERMINAL SCREW

In order to wrap a cable wire around a screw terminal so that you achieve a proper, durable connection, you need to work with first a wire stripper (or knife), then a pair of long-nosed pliers, and finally a screwdriver.

1. Strip about one inch of insulation from the end of the wire. If you are using wire strippers, the insulation should slide off the wire easily and cleanly, but check the metal anyway and scrape away any insulation stuck to it with the *back* side of a knife blade.

If you are using a knife to strip the insulation, cut around the insulation, holding the knife at a 60° angle away from the end of the wire, so that the end of the insulation is tapered. Be careful not to cut so deeply that you nick the metal. Then pull the insulation off the wire with your fingers or a pair of pliers, and clean the wire of any remaining insulation.

2. Using long-nosed pliers, bend the bared end of the wire into a tight loop that is just large

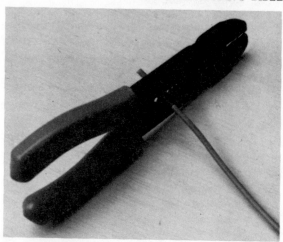

Fig. 46. A multipurpose tool will strip wire cleanly without scraping the copper.

enough to go around the terminal screw. When you are finished bending, the loop should look like a question mark.

MAKING A TERMINAL SCREW CONNECTION

1. Back the terminal screw out of its hole until the "upset" prevents it from coming out any farther.

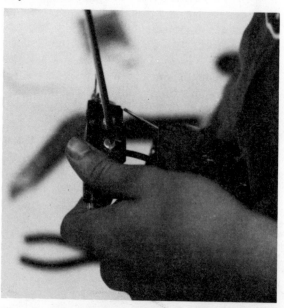

Fig. 47. The wire loop must be placed around the terminal in a clockwise direction so that when the screw is tightened it will tend to draw the wire under its head, rather than push it away.

2. Place the bent wire end around the terminal so that the loop is going in a clockwise direction. It is important that the loop be clockwise so that when the screw is tightened (in a clockwise direction) the wire will be drawn beneath the head of the terminal, rather than squeezed out from under it.

3. If at all possible, pinch the end of the loop around the threads of the terminal with your pliers.

4. Tighten the terminal screw as tightly as you can get it. There should not be more than ¼" of bared wire between the edge of the screwhead and the end of the insulation on the wire, and no part of the loop should be visible at all.

SPLICING WIRE

It is hazardous to splice wires at any point in a home wiring system, except where the splices can be housed in an electrical box. The NEC accepts two ways of splicing wires: either by twisting the wire ends together and taping the connection, or by use of a screw-on connector commonly known as a wire nut. The NEC does not allow any solder to be used in service equipment, in the ground wire, or any grounding wires, on the basis that an unusually high surge of electricity might well melt the solder and open the connection all together. Soldering, therefore, is no longer used in any part of house wire systems. At any rate, no matter how you make it, a splice must meet three criteria:

1. It must be mechanically as strong as an unbroken conductor.

2. It must be able to conduct electricity as if it were a continuous piece of wire.

3. The insulation around it must be equal to the original insulation that covered the wires before you made the splice.

SOLDERLESS CONNECTORS

There are several types of solderless connectors available, but the most common ones are wire nuts, with a copper spring coiled inside a plastic thimble. The types that do not have a plastic shell must be wrapped in electrician's tape, which is an added step that does not occur with the shelled versions. You can purchase wire

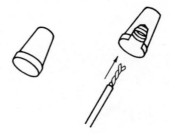

Fig. 48. Cutaway view of a wire nut.

nuts in a variety of sizes for use with different wire gauges, or to fit a number of wires coming together in the same splice. Here's how they work:

1. Remove between 2" and 3" of insulation from each of the wires to be spliced. If you are connecting several wires that are the same diameter with a smaller-sized wire, take a little more insulation off the small wire so that it can be wrapped around the outside of the bigger wires.

2. Twist the ends of the wires together by holding them next to each other and then rotating them clockwise. If you are working with cord wires, you can do this with your fingers, but if you are splicing circuit cable wires you will need a pair of adjustable-end pliers. Twist the wires as tightly together as you can get them. Then yank on them slightly to make sure they stay together.

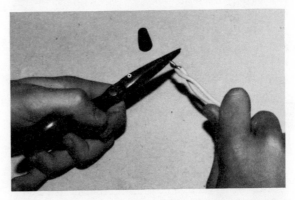

Fig. 49. You can actually splice two wires together with a wire nut merely by twisting the nut down over the wires, but if possible, twist them together first.

3. Screw the wire nut down over the ends of your splice, turning it clockwise until the nut will not go any farther. At this point the bottom of the wire nut *must* cover all bared wire. It should,

in fact, cover a fraction of an inch of insulation at the base of the splice. Tug at the wires to be certain the splice is secure.

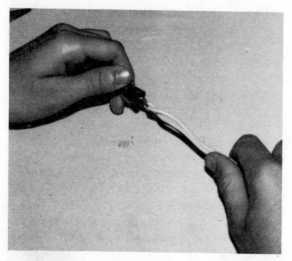

Fig. 50. Rotate the wire nut down on the wire ends until the base of the nut covers the insulation behind the bared wire ends.

4. If you have used a solderless connector that does not have an insulating shell, wrap the wires and the connector with plastic electrician's tape.

Wire nuts can be used to hold stranded as well as solid wires. Before you wrap the strands together, twist the strands on each wire tightly in a clockwise direction. Then twist all of the wires together and cap them with a wire nut.

If you install a wire nut over any connection and find that it does not cover the insulation at the base of the splice, remove the wire nut by rotating it counterclockwise and snip off the ends of the spliced wires with wire cutters, then apply the wire nut again.

TAPING A SPLICE

There are several ways of taping a splice. Until several years ago, in order to make a proper taped splice you had to wrap the bare wires with a rubber tape known as splicing compound and then cover that with friction tape to protect the rubber. Now we have plastic electrician's tape, which is mechanically strong and has a high insulating value, so that a comparatively thin layer will sufficiently insulate and hold the wires together. No longer do your taped splices need to

look as though you have bandaged somebody's elbow.

When applying tape over a splice, start on the insulation at one end of the bared wire and wind the tape spirally around the insulation and the splice. When you reach the insulation at the other end of the splice, wind the tape back again, overlapping each turn slightly. Keep the tape stretched as you wrap it and continue wrapping back and forth until you have built up enough layers of tape to equal the thickness of the wire insulation.

You can wrap a splice at the ends of several wires as if it were a cone. You can also bend the splice downward between the wires and then tape the splice between the wires.

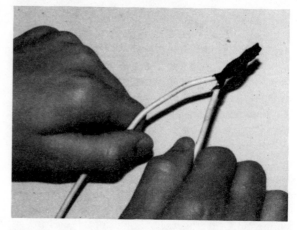

Fig. 51. Wrap electrician's tape spirally around the ends of your spliced wire. The wire must be twisted together to form a secure splice.

TAPPING SOLID WIRE

On occasion, you may need to tap into a continuous wire with a second wire. The simplest approach to tapping is to cut the wire to be tapped and strip both its ends, then strip the end of the tapping wire and join all three wire ends with a wire nut. If this is not practical, you can make a tap using plastic electrician's tape:

1. Strip about 3″ of insulation from the continuous wire at the point you wish to make the tap. You cannot use a wire stripper for this but will have to cut around the insulation at both ends with a knife and then slit the covering lengthwise between your vertical cuts.

2. Strip about 3″ of insulation from the end of the tapping wire.

3. Wind the end of the tapping wire around the bared portion of the continuous wire.

4. Cover the tap with plastic electrician's tape, beginning on the insulation at one end of the continuous wire, and ending at the other end of the continuous wire. Continue wrapping the tap until the thickness of the tape equals the thickness of the insulation on the wires.

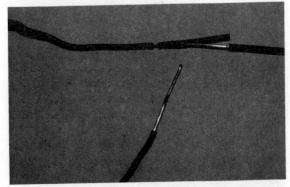

Fig. 52. Strip a portion of the wire to be tapped in the middle.

Fig. 53. Wrap the end of the tapping wire tightly around the bared section of wire.

Fig. 54. Wrap all of the bared wire with electrician's tape.

HOW TO TEST FOR GROUNDING

The purpose of grounding is safety. Not all home wiring systems are completely, or even partially, grounded. You can test a receptacle's grounding by using a neon test lamp. If the receptacle is one of the older types, it has two slots that are equally long. You have no way of telling which hole is the hot line unless you take off the faceplate and look (don't touch) at which side of the receptacle the black wire is connected to. If the receptacle is a modern, 3-hole unit, its slots are different in length. The longer (or wider) slot is supposed to be connected to the grounded (neutral) white wire of the branch circuit. The shorter (narrower) slot is supposed to be connected to the black hot line. They may not be hooked up that way because the electrician who installed the receptacle may have made a mistake. In any event, the way you test the receptacle for grounding is this:

1. Be sure the center screw holding the faceplate over the receptacle is not painted. If it is, remove the faceplate or scrape the paint off the screw.

Fig. 55. One of the slots in a 3-hole receptacle is shorter than the other. The short slot is supposed to be connected to the black hot line.

2. Insert one probe of the hot-line tester in the narrow slot, which should be the slot that is on

the same side of the receptacle that the black hot line is attached to.

3. Touch the other probe of the tester to the center holding screw or the edge of the metal electrical box. If the light goes on, the circuit is grounded. If it does not go on, the grounded wire is either broken or disconnected.

If you want to test the receptacle for continuity, insert the probes of your hot-line tester into the slots of the receptacle. If the light goes on, the receptacle is "live."

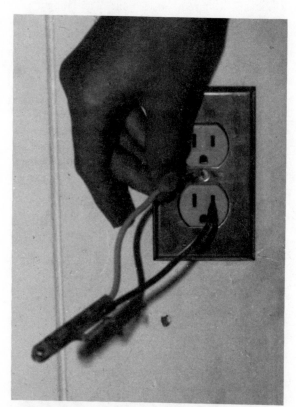

Fig. 56. Insert one probe of a hot-line tester in the narrow slot of the receptacle and touch the other probe to the edge of the electrical box or the screw holding the faceplate (if it is unpainted). The tester light will go on if the circuit is grounded.

WORKING WITH ARMORED CABLE

Armored cable these days is sold with a bare grounding wire inside its flexible metal sheath. The wires are insulated and then wrapped in heavy paper as further protection. From a working standpoint, armored cable is an awkward material to deal with, primarily because it cannot be bent into a tight radius. It does, however, provide a continuous grounding system via its metal armor, so even if you somehow fail to use the grounding wire provided in each cable, your electrical system will still offer a measure of protection against shock hazards.

CUTTING ARMORED CABLE

1. Hold the cable as steady as you can and angle the blade of a hacksaw *across* the diagonal spiral of the armor (Fig. 57). Do not cut with the armor's spiral, but against it. Normally, when an electrician is stripping the armor off a cable that will attach to a receptacle or switch, about 10″ or 12″ of armor is removed so that there is plenty of wire inside the electrical box to work with.

2. Saw through the armor on one side, then rotate the cable and continue sawing until you have cut all the way around the armor. Be very careful not to cut so deeply that you touch the wire insulation.

3. Pull the severed portion of armor off the wires.

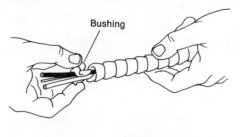

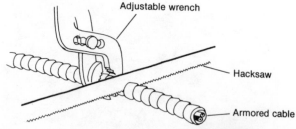

Fig. 57. The sharp edges of a cut must be covered with a red-fiber bushing to protect the wires from being damaged (top), after the armored cable has been cut with a hacksaw (bottom).

INSTALLING ARMORED CABLE IN ELECTRICAL BOXES

1. The Code requires that all severed ends of armored cable be covered by a fiber bushing to protect the wires from being cut by the sharp metal of the armor. The bushings are a bright red, and they come with the cable when you buy it. Each bushing is split so that it can be opened around the wires and then pushed down inside the armor. There is a lip on the top of the bushing that fits over the edge of the armor. Don't skip the procedure of putting on bushings; building inspectors specifically look for them when they are inspecting electrical work and they will reject any connection that is missing its bushing.

2. Bend the bare grounding wire coming out of the cable back over the armor and wrap it tightly around the metal.

3. Slide the wires through the cable side of an armored cable connector. The cable side has no threads, but there is a setscrew on the connector (Fig. 58). Push the connector down over the armor as far as it will go and tighten the screw. Tug on the connector to be sure it is securely fastened to the cable.

4. The threaded side of the connector has a flanged locknut on it. Take the locknut off and insert the connector's threaded end through the appropriate knockout in an electrical box.

5. Hand-tighten the locknut on the threaded side of the connector. Then place the blade of a screwdriver against any of the flanges on the locknut and tap the screwdriver until the nut bites into the metal of the box. The cable wires are now ready to be stripped and connected to whatever device the electrical box is to contain.

Some electrical boxes come with a screw-tightened clamp inside the box to hold the primary cables entering and leaving the unit. These are used by inserting the cable through the knockout behind the clamp until the bushing grounds against the face of the clamp. The clamp is then tightened with a setscrew.

WORKING WITH NONMETALLIC-SHEATHED CABLE

Nonmetallic-sheathed cable is far more flexible than armored cable but has the disadvantage of requiring that its bare grounding wire be con-

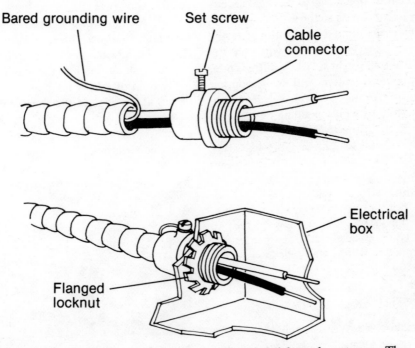

Fig. 58. Insert the end of the cable in its clamp and tighten the setscrew. The clamp is held in the box knockout by a flanged locknut.

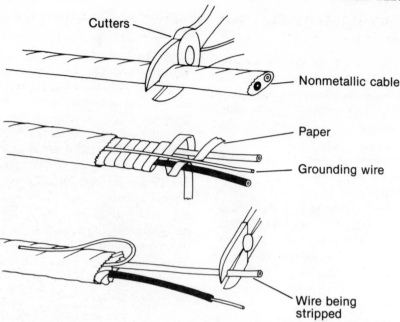

Cutters

Nonmetallic cable

Paper

Grounding wire

Wire being
stripped

Fig. 59. Cut through the nonmetallic sheathing with cutters or a knife.
Remove the protective paper around the cable wires, then strip insulation
from the ends of the wires.

nected to metal before there is any kind of grounding system in operation to protect the circuit. The way to cut it is as follows:

1. Carefully cut around the sheathing at the point that you want to bare the wires in the cable (Fig. 59). You ought to use a utility knife, but any sharp knife or cutters will do. If you are installing the cable in an electrical box, leave between 10″ and 12″ of wire extending beyond the sheathing.

2. Carefully slice lengthwise through the sheathing from your first cut to the end of the cable and peel the sheathing away from the wires. Alternatively, you can make the horizontal cut first, then peel back and snip off the sheathing. The advantage to this approach is that you do not risk cutting into the wire insulation.

3. Snip off all of the paper packed around the wires at the edge of the sheathing. You now have a white wire, one or two black wires, and a bare copper wire sticking out of the end of the nonmetallic sheathing.

Fig. 60. Be careful when slitting nonmetallic-sheathed cable. Don't cut into the wires.

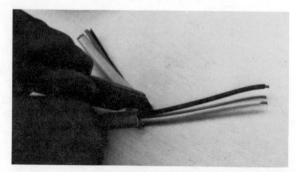

Fig. 61. Cut off the protective paper and the sheathing.

INSTALLING NONMETALLIC-SHEATHED CABLE IN ELECTRICAL BOXES

The clamps used to hold nonmetallic-sheathed cable in a metal electrical box consists of a threaded connector which has a removable yoke on its cable side. The yoke is held in place with a pair of screws that tighten it against the cable (Fig. 62).

1. Insert the threaded end of the clamp through a knockout in the electrical box.

2. Put on the flanged locknut and hand-tighten it against the side of the box. Be sure the locknut bites into the metal by hammering a screwdriver against the flanges.

3. Loosen the lock screws holding the clamp yoke and push the wires through the connector until the sheathing is against the inside end of the clamp yoke.

4. Tighten the screws on both sides of the yoke until the cable is secure. Pull at the cable to be certain it is tightly in place.

Some electrical boxes include a nonmetallic-sheathed clamp, which amounts to a yoke that is tightened down over the cable with a single screw. Insert the cable through the clamp and tighten the yoke with a screwdriver.

GROUNDING CABLE TO ELECTRICAL BOXES

You have several options for grounding the bared wires in any cable. You can install a screw in any of the tapped holes in the sides of the box

Fig. 63. One way to ground the wires in an electrical box is to clip them to the edge of the box with a special grounding clip.

and wrap the wire around the screw, then tighten it. You can also use special solderless connectors that hold the grounding wire and clamp over the edge of the metal box. You are allowed to splice all of the grounding wires entering a box together with a wiring nut and wrap only one of them around a single screw set in the back of the box.

WORKING WITH CONDUIT

Most communities require that any cable installed outdoors be protected by conduit and weatherproof electrical boxes. Some conduit comes in a thin-walled variety that is rigid but thin enough to be bent using a hickey bending tool. The thicker, more rigid type of conduit is more like a waterpipe and must be assembled to its electrical boxes with threaded couplings and fittings.

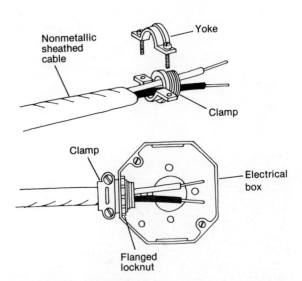

Fig. 62. Clamps used to hold nonmetallic-sheathed cable have a yoke that is secured against the cable with screws. The clamp is installed in an electrical box with a locknut.

BENDING CONDUIT

The hickey is a curved, grooved device with a long handle. Fit the curved channel of the hickey over the end of the pipe to be bent and lock the pipe in place with the hook attached to one end of the curve on the hickey (Fig. 64). Bend the pipe by standing on it and pulling the hickey to-

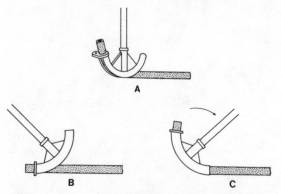

Fig. 64. How to bend conduit with a hickey. *A* shows the function of the tool from underneath, *B* and *C* show the proper motion to use.

ward you, moving it along the pipe if you want a more gradual bend. Check the bend from time to time to be certain you have the correct angle you need.

PULLING WIRES THROUGH CONDUIT

When you have assembled conduit between junction boxes, wires must be pulled through the conduit. Under no circumstances does the NEC allow any splices inside any conduit. The Code also requires that a ½″ diameter conduit may contain no more than four ⌗14 or three ⌗12 wires. A ¾″ conduit may contain no more than six ⌗14 or five ⌗12 wires. In order to pull the wires through the conduit, insert an electrician's snake or fish tape, and attach the pipe and the end of the wires to the tape. Pull the tape and wires through the conduit. Conduit is manufactured with an unusually smooth inside finish that allows wires to slide through them with ease.

CHAPTER FIVE

Home Wiring Projects

Home wiring projects involve any work done to a house electrical system, including replacing receptacles and switches, installing light fixtures indoors and outdoors—any work that involves working directly with branch circuits that already exist in the house.

The branch circuits in your home are presumably all live, and electricity is coming as far as, if not into, all of the receptacles, switches, and fixtures attached to the circuit cables. Without exception, before you begin any work on any part of your home electrical system, always disconnect the source of power by removing the branch line(s) fuse or turning the circuit breaker to its OFF position.

RECEPTACLE WORK

Receptacles or outlets can, for no discernible reason, stop functioning. Since they are normally fully encased pieces of equipment, there is little or no percentage in trying to repair them, particularly since a replacement unit costs so little (about one or two dollars). Ideally, if you have old 2-slot receptacles in your home, you would replace all of them with modern, 3-hole units that can also be grounded to their boxes. The procedure for replacing any receptacle, no matter how it is designed, is pretty much the same, and is in fact illustrated on the back of the package that contains it.

HOW TO REPLACE A RECEPTACLE

1. Turn off the power.
2. Remove the screw or screws holding the faceplate (outlet plate) over the front of the receptacle (Fig. 65).

3. There are mounting screws at the top and bottom of the receptacle that hold the unit to its electrical box. Remove them.

4. Pull the unit straight out of its box. The cable wires are thick and stiff, so they will resist you somewhat, but they *will* bend, and allow you to get the receptacle as much as 6″ or so out of the box.

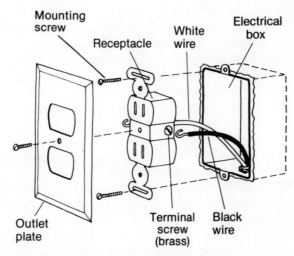

Fig. 65. How a receptacle is wired to its box (grounding wire removed for clarity).

5. Loosen the terminal screws on the sides of the receptacle that hold the black and white cable wires and unhook the wires. If there is a grounding wire attached to the receptacle, its terminal is at one end of the unit and is painted green. Remove the grounding wire, too.

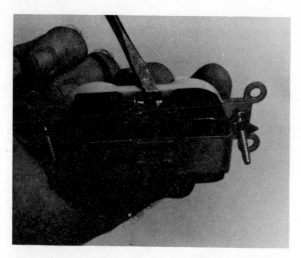

Fig. 66. The metal connecting tab can be broken with pliers or by twisting it with a screwdriver.

Fig. 67. The green wire attached to the green grounding terminal (opposite screwdriver) attached to all modern receptacles.

6. New receptacles have four terminal screws, two brass on one side and two silver-colored on the other. Look closely at the terminals and you will see a small metal tab between them that can be broken off to separate the outlets. Under most circumstances, you will leave the tab alone, allowing the connection of the black cable wire to either brass terminal and the white wire to either silver-colored terminal, which brings power into both outlets in the receptacle. You would break the tabs, and thereby isolate the outlets, only if you wanted the outlets to be separately powered.

7. When you have connected the white cable wire to one of the silver-colored terminals on the new receptacle, and the black wire to one of the brass terminals, you must then deal with the green grounding terminal. You can purchase short lengths of green wire at most stores that sell electrical equipment. The wire is often sold with a terminal screw attached to one end of the wire that will fit into any of the tapped holes in the sides of any electrical box. So all you have to do is thread the terminal screw into a hole in the box, then connect the free end of the green wire to the green terminal screw on the receptacle. If you do not have a prepared grounding wire, you can strip a few inches of bared wire out of a cable (any cable wire you wish, actually) and connect it between the green terminal screw and the box. Its connection to the box can be with either a screw or a special grounding clip; it can

be attached to the box separately, or to the same connection that holds the cable grounding wires.

8. With the wires all connected, push the receptacle toward the box until its top and bottom flanges are against the front edges of the box. The stiffness of the cable wires will resist you somewhat, but again, they *do* bend and you can force them to do so without breaking them.

9. Secure the receptacle to its box by tightening the top and bottom screws that came with it

Fig. 68. Push the wires into the box and secure the receptacle with its holding screws.

in its holding flanges. Then replace the faceplate and secure it with its screw.

10. Turn on the current interrupter and test the receptacle. If the fuse blows or the circuit breaker snaps to OFF as soon as you activate the interrupter, you have a short circuit in the electrical box, most likely because one of the wires is touching something it shouldn't. Turn off the current, pull out the receptacle, and double-check all your connections.

WIRING A TWO-CIRCUIT DUPLEX RECEPTACLE

There are occasions when you might need to split the current entering a duplex receptacle. In a kitchen, for example, you might want the outlets in a series of small appliance receptacles to be on separate circuits. Or you might want to have one outlet in your living room always hot, so that you can plug a vacuum cleaner, clock, or radio into it but still be able to control a lamp plugged in the other outlet with a wall switch.

When you want to split the power entering a receptacle so that one of the outlets is controlled by a switch, first break off the tab between the brass terminals on the unit. Do *not* break off the tab between the silver-colored terminals. Then connect the white wire from the cable to either of the silver-colored screws. Splice the black wire from the cable to the black wire from the switch and a short length of wire. Connect the short wire to the brass terminal next to whichever outlet you want to remain hot. The white wire from the switch should be taped or painted black in accordance with NEC rulings and connected to the unused brass screw on the outlet to be controlled by the switch.

If you are separating the incoming power in a receptacle that is not switched, break off the tabs between *both* the brass and the silver-colored terminals. Then connect the white and black wires from one cable to one of the outlets and the wires from the other branch cable to the terminals on either side of the other outlet.

REPLACING SWITCHES

Switches represent the only instance in house wiring when a white wire may be used as a hot line. The reason for this is based on practicality. Cables are manufactured with one black and one white wire, or with a black, a red, and a white wire. You need to run two wires from a switch to whatever it is controlling, but neither of the wires is a return line. They are both parts of the hot side of the circuit. So it is common practice in the electrical industry to use a 2- or 3-wire cable when connecting switches, but when the white wire in the cable is used as a hot wire, its ends are painted black or wrapped with black tape to signify that the line is hot.

The one concept to understand about switches is that they are connected to the hot side of a circuit. That is, they stand between the incoming current and the circuit load. Consequently, the wire leading into a switch is hot, and the wire leading away from it is also hot. If there were a white grounded wire also connected to the switch, it would form a return line to the source of electricity before the current could reach the circuit load. Since that would inhibit the operation of the light or device being switched, the white return line can only go from the device back to the source of voltage.

Although replacing an old switch with a new one is normally a simple procedure, you can open some switch boxes and discover a maze of wires going in all directions, all looking as if they were black. In this case, be very careful which wires you take off the old switch and which terminals on the new unit you put them on. In situations of extreme confusion, take one wire off the old switch and put it on its corresponding terminal on the new unit, then take the next wire off the old switch and put that one on the new switch.

If you make a mistake in where you put the wires on your replacement switch, there will be no bolts of lightning shooting out of the device when you turn on the power. There will, in fact, be nothing. The light simply will not work and you will have to try different wires on different terminals until it does.

In any case, the steps to take when replacing a switch are as follows:

1. When the current interrupter has been removed or switched off, take off the switch faceplate.

2. Look inside the switch box. If there are

more wires in there than you expected (two plus a grounding wire), locate the other circuits entering the box and turn them off, or shut down all of the power entering the building. Switch boxes are sometimes used as junction boxes, and more than one cable may be crossing through them. If the switch you are repairing is ganged with other switches, the box will be two or three times larger than normal, and there could be three or four times as many wires entering it as well.

3. Loosen the setscrews at the top and bottom of the switch and pull the switch out of the box. If there is more than one switch in the box, they may be interconnected with a short jumper wire. To get at the interconnection, you may have to pull more than one switch out of the box.

4. Remove one wire from the old switch and put it on the corresponding terminal on the new switch.

5. Remove the second wire from the old switch and put it on the corresponding terminal on the new switch.

6. Transfer the grounding wire from the green terminal on the old switch to the green terminal on the new switch. If the old switch is not grounded, ground the new one to the electrical box (see page 43).

7. Push the new switch back against the face of the box and secure it with its setscrews.

8. Install the faceplate over the switch and turn on the circuit interrupter. If the switch does not work, you have put the wrong wires on the terminals.

COMBINATION UNITS

Designed to fit into any electrical box, combination units may be three outlets, a switch and two outlets, a switch, outlet and night-light, or some other grouping of two or more devices. As a series of devices, the combination unit tends to require more wires than you can comfortably get into a standard switch box, so use a *deep 4″* electrical box. Like the outlets discussed earlier, most of the combination units you will encounter in electrical and hardware stores have their terminals connected by metal tabs between them, giving you the option of putting all or only one

or two of the three units on the same branch circuit. In some models, however, each unit is separated from the others, so if you want to put all three devices on the same circuit, a jumper wire must be connected between the terminals.

The procedure for installing a combination unit is identical to replacing a receptacle. You connect the white cable wires to the silver-colored terminals and the black wires to the brass terminals. If you have separate cables entering the electrical box, the appropriate cable wires are attached to each device. The outlet and night-light will receive black and white wires. The switch will probably also have a black and white wire, but the white conductor must be colored black to signify that it is being used as a hot line.

If you wish to connect the three devices in a combination unit to the same branch circuit and they do not have metal links between their terminals, you need black and white jumper wires to bridge the brass and silver-colored terminals. These jumpers can actually be partially stripped extensions of the cable wires, or they can be separate wires (Fig. 69). If the jumper is a separate wire, it will have to be doubled on one of the terminals, which is the easiest thing to do even though the NEC prohibits more than one wire on any given terminal. The procedure for making and installing jumper wires and also satisfying Code safety requirements is this:

1. Cut a 12″ length of black cable wire and a similar length of white cable wire. The wire gauge should be the same as the gauge used in the branch circuit feeding the combination unit.

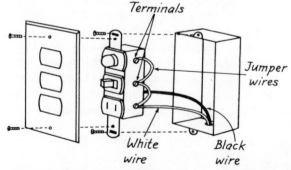

Fig. 69. If you leave a foot or so of wire from the cable after you connect it to the first terminal, you can use this extra length to act as a jumper between components in a combination unit.

Fig. 70. A jumper wire connecting the two parts of a combination switch and receptacle.

2. Strip between 1″ and 2″ from each end of both wires. Also remove about 2″ of insulation from the center of each wire.

3. Connect the black wire to the top brass terminal on the combination unit. Then bend the wire so that you can loop the uninsulated center portion of the wire around the middle terminal.

4. Repeat step ⌗3 with the white wire and the silver-colored terminals. You now have two of the devices connected to each other.

5. Cut a 6″ length of black cable wire and a 6″ length of white wire. Strip 2″ from both ends of each wire.

6. Twist the end of the black jumper cable together with the black circuit cable wire and one end of the 6″ length of black wire. Secure your splice with a wire nut.

7. Repeat step ⌗6 with the white wires.

8. Connect the free end of the 6″ black wire to the black brass terminal at the bottom of the combination unit. Connect the 6″ white wire to the silver-colored terminal on the bottom device.

9. Install a grounding wire between the combination unit and the electrical box.

BACK-WIRED DEVICES

Some switches and receptacles are manufactured without screw terminals, but with small holes in their backs to receive the ends of cable wire. Many of the newest switches and receptacles on the market have both the holes and terminals.

There is a marker (in reality a groove) cut in the back of the device that tells you how long the bared end of the cable wire should be, and once you have stripped off that much insulation (about ½″) the wire is merely pushed into its proper hole, where it is gripped by a spring clamp. To disconnect a wire from its hole, you have to push the blade of a screwdriver into a slot next to the hole (which releases the clamp) and simultaneously pull the wire free. Back-wired devices are a little easier to install, since the cable wire never has to be looped around a screw, but be aware that some building codes do not favor back-wired devices and your local building inspector may reject them, even though they are UL-listed.

Fig. 71. The marker on the back of a back-wired switch or receptacle tells you exactly how much wire to bare.

Fig. 72. The bared wire is pushed into the proper hole and is locked in place by a spring clip inside the unit.

Fig. 73. The wire is released by pushing down on the spring with a screwdriver blade.

INSTALLING DIMMER SWITCHES

One of the best ways you can reduce your use of electricity each month is to control many or all of your overhead fixtures with dimmer switches. The dimmer, no matter which of the many types you install, is designed to reduce the amount of electricity going to a light bulb, causing it to burn with less brightness than it would normally produce. In dining rooms, family rooms, and living rooms, in particular, a dimmer controlling the overhead lighting can conserve tremendous amounts of electricity.

Fig. 74. Dimmer switches are made to fit into any standard switch box.

Dimmers are manufactured in a variety of designs and each is rated as to the maximum wattage that it can handle. The least-expensive, simplest models can usually control up to 300 watts, while others can handle as much as 1000 watts. Be certain that the dimmer you install is capable of handling the total number of watts in the bulbs it must control. You cannot interchange dimmers between ordinary incandescent and fluorescent lighting, but there are separate units for use with fluorescent lighting, just as there are separate units that can be installed to handle 3-way bulbs.

Dimmers are all engineered to fit into a standard switch box, and they are installed exactly as if they were any toggle switch. You connect one black wire to the black cable wire and the other dimmer wire to the white cable wire colored black. Some dimmers have screw or black-wire

Fig. 75. Some dimmers have leads, or wires attached to them which are spliced to the wires in the electrical box with wire nuts.

terminals, others come with two wires sticking out of them, which are spliced to the cable wires with wire nuts. The unit should also have a way of being grounded to the electrical box via a green grounding wire. Don't neglect to ground the dimmer, as you would any switch.

WIRING SWITCHES, FIXTURES, AND RECEPTACLES

Over the years, in millions of installations, electricians have devised uncounted numbers of ways of connecting switches, fixtures, and receptacles to each other. Though the following wiring arrangements are by no means complete, they do represent some of the more common wiring schemes:

FIXTURE AND ONE SWITCH

In this arrangement, the 2-wire circuit cable enters the switch box (Fig. 76). A second 2-wire cable goes from the switch box to the fixture box. The bare grounding wires from the two ca-

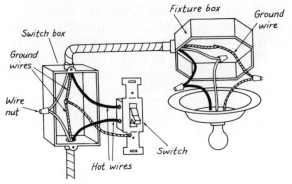

Fig. 76. Fixture and one switch.

bles in the switch box are connected to the grounding wire (bare or green) from the switch and all three are connected to the switch box. The black wire from the circuit cable connects to one terminal on the switch. The black wire from the cable leading to the fixture box is connected to the other terminal on the switch. The two white wires in the cables are spliced together with a wire nut.

The bare cable wire in the fixture box is spliced to the green grounding wire from the light fixture. The black cable wire from the cable is spliced to the black fixture wire. The white cable wire is spliced to the white fixture wire.

SWITCH THROUGH A JUNCTION BOX

This time the circuit cable is running through the junction box and the switch must be tapped into it in order to control a light somewhere at the end of the circuit (Fig. 77). A 2-wire cable is run from the junction box to the switch box. The bare cable wire is united with the grounding wire from the switch and they are both attached to the switch box. The black wire from the cable goes to one of the switch terminals and the white

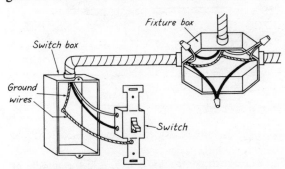

Fig. 77. Switch through a junction box.

wire to the other. The white wire must be painted or taped black.

ADDING AN OUTLET TO AN OUTLET

The branch circuit terminates at an outlet, but you want to add another receptacle to the circuit (Fig. 78). The existing outlet has a single 2-wire cable entering it and its wires take up two of the terminals on the receptacle. Run a 2-wire cable between the outlet boxes. In the old box, the bared grounding wire of the new cable is attached to the grounding wires from the receptacle and the old cable, and all three are attached to the electrical box. The white wire from the new cable connects to the remaining silver-colored terminal on the receptacle. The new cable's black wire connects to the remaining brass terminal on the receptacle (assuming the tabs are still attached between the terminals).

In the new electrical box, the new cable's bare wire is spliced to the grounding wire from the new receptacle and both are attached to the electrical box. The white wire from the cable is attached to a silver-colored terminal on the receptacle; the black wire connects to a brass terminal.

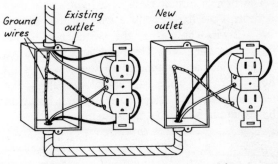

Fig. 78. Adding an outlet to an outlet.

TWO-GANG SWITCHES CONTROLLING DIFFERENT LIGHTS

These two switches are physically next to each other in the same box, which has been constructed by removing one side of each box and hooking the boxes together (Fig. 79). The switches are powered by separate branch circuits that enter (and leave their boxes). In this situation, you treat each switch as if it were by itself.

The white wires from the two ends of the same cable are spliced with a wire nut. The black wires from the cable are attached to the switch terminals. The bare grounding wires in the cable ends are connected to the grounding wire from the switch and all three wires are attached to the electrical box.

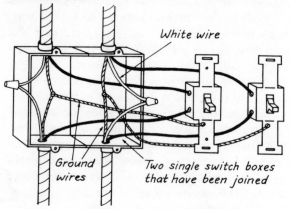

Fig. 79. Two-gang switches controlling different lights.

SWITCH AND FIXTURE AT THE END OF A BRANCH LINE

The branch circuit ends at a light. You want to control the fixture with a wall switch, so you run a 2-wire cable from the switch box to the fixture box (Fig. 80). The black and white wires of the cables are connected to the terminals on the switch. The white wire is colored black; the bare cable wire is bonded to the switch box with the grounding wire from the switch.

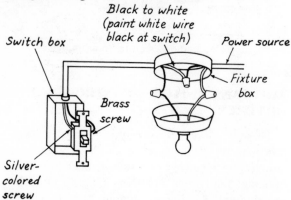

Fig. 80. Switch and fixture at the end of a branch line (grounding wires removed for clarity).

The black cable wire enters the fixture box and connects to the black wire from the fixture, while its white wire (colored black) is spliced to the black wire from the branch cable. The white wire in the branch cable is spliced to the white wire from the fixture. The grounding wires from the fixture, the circuit cable, and the cable linking the switch and the light are all secured to the fixture box.

SWITCH CONTROLLING A FIXTURE IN THE MIDDLE OF A CIRCUIT

When the fixture is in the middle of a branch circuit, two cables enter the fixture box (Fig. 81). By adding a controlling switch, you end up with three cables and a box full of wires. The cable that links the switch to the fixture has two wires, both of them connected to terminals on the switch. The white wire must be painted or taped black. The cable grounding wire can be united with the switch grounding wire and then both bare wires are connected to the switch box.

At the fixture box, the black wire from the switch is spliced to the black fixture wire. The white wire from the switch is colored black and is spliced with the two black wires from the branch cables. All of the white wires from the circuit cables and the fixture are spliced together. The bare grounding wires from the circuit cables, the switch cable, and the fixture are attached to the fixture box.

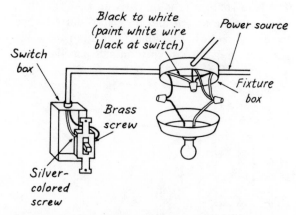

Fig. 81. Switch controlling a fixture in the middle of a circuit (grounding wires removed for clarity).

TWO FIXTURES CONTROLLED BY DIFFERENT SWITCHES

You have two overhead lights on the same branch circuit (Fig. 82). The unit at the end of the circuit is controlled by a pull chain that is actually a part of the fixture. The other overhead light, which is in the middle of the circuit, is controlled by a wall switch. If the power reaches the fixtures by first entering the switch box, it must then continue to both lights. So there is a 3-wire cable from the switch to the first light, and a 2-wire cable between the two fixtures.

The white wire from the circuit cable is spliced with the white wire from the 3-wire cable, inside the switch box. The two black lines from the cables are attached to one of the switch terminals, preferably via a short jumper wire. The red wire is attached to the other switch terminal. All of the cable and switch grounding wires are attached to the box.

The 3-wire cable enters one side of the first fixture box, but a 2-wire cable leaves it to connect with the second fixture. The red wire in the 3-wire cable is spliced to the black fixture wire. All of the white wires from the two cables and the fixture are spliced together. The two black cable wires are spliced and the cable and fixture grounding wires are attached to the box.

The 2-wire cable entering the last fixture is a straight hookup. The white wire from the cable is spliced to the fixture white wire and the two black wires are connected. The cable and fixture grounding wires are attached to the electrical box.

ADDING AN OUTLET AND SWITCH TO A FIXTURE

Assuming that a branch circuit ends at an overhead light and you want to add not only a switch but also an outlet, you must, in effect, extend the branch circuit through the switch and end it at the outlet (Fig. 83). To do this, you need to run a 3-wire cable from the fixture to the switch box and a 2-wire cable from the switch to the receptacle. In the fixture box you now have a 2-wire cable which is bringing power to the light, a 3-wire cable which is continuing the power to the switch and receptacle, and a light fixture. Counting the three grounding wires, what you really have is an upside-down bowl of copper spaghetti. All of the white wires are spliced together. All of the grounding wires are attached to the fixture box. The two black wires from the cables are spliced together. The red wire from the 3-wire cable is spliced to the black wire from the fixture.

The switch box contains the 3-wire cable and a 2-wire cable leading to the receptacle, plus the switch. The white cable wires are spliced together. The bare grounding wires and the switch grounding wire are connected to the box. The red wire is attached to one of the switch terminals. The two black cable wires are spliced to a third jumper wire, which in turn is connected to the other switch terminal.

The receptacle is an anticlimax. The white wire of the 2-wire cable from the switch box is

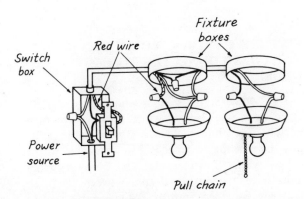

Fig. 82. Two fixtures controlled by different switches (grounding wires removed for clarity).

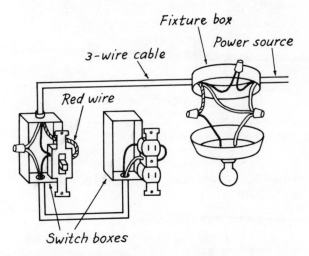

Fig. 83. Adding an outlet and switch to a fixture (grounding wires removed for clarity).

attached to the silver-colored terminal on the receptacle. The black wire from the cable attaches to the brass terminal on the receptacle. The bare cable wire and the receptacle grounding wire are bonded to the electrical box.

ADDING ON TO A JUNCTION BOX

The wiring hookups for adding a light fixture or an outlet are identical if you are coming from a junction box (Fig. 84). You must run a 2-wire cable from the junction box to the light or receptacle position. The black wire of the cable connects to the black wire on the fixture or the brass terminal on a receptacle. The white cable wire attaches to the silver-colored terminal on the receptacle, or to the white wire on a fixture. The bare wire from the cable and the grounding wire from either the light or the receptacle are connected to the electrical box.

Meanwhile, back at the junction box the white wire from the 2-wire cable is spliced together with all the other white wires in the box. And the black wire from the cable is spliced to all the other black wires in the box. The cable grounding wire is attached to the box.

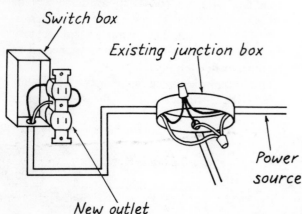

Fig. 84. Adding on to a junction box (grounding wires removed for clarity).

ADDING A SWITCH AND OUTLET TO A FIXTURE

It requires very little extra work to add an outlet next to a switch you are installing, and for that very little work, you can gain the added versatility of another outlet in the room (Fig. 85).

This is particularly useful in bathrooms and laundry rooms, but it can be done anywhere. You can reduce your work by putting the switch and receptacle in the same box, which you construct by attaching two switch boxes. Pull a 3-wire cable from the fixture to the switch side of the box. The black wire from the fixture is spliced to the red cable wire, and the black cable wires are connected to each other. The three white wires (from the two cables and the fixture) are also spliced together. The grounding wires are all attached to the box.

The red wire in the 3-wire cable is attached to one terminal on the switch. Carry the black wire from the cable to the other switch terminal, but don't cut it off there. Remove 2″ of insulation in the middle of the wire and wrap the bared wire around the terminal. Then continue the wire to the brass terminal on the receptacle. The white cable wire is attached to the silver-colored terminal on the receptacle. The cable, switch, and receptacle grounding wires all connect to the box.

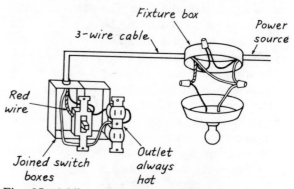

Fig. 85. Adding a switch and outlet to a fixture (grounding wires removed for clarity).

FIXTURE CONTROLLED BY DIFFERENT SWITCHES CONNECTED TO EACH OTHER

The switches, no matter where they are located, must be 3-way switches which are connected to each other via a 3-wire cable (Fig. 86). One of the switches is also connected to the fixture with a 2-wire cable. The 3-way switches must have a complete circuit between them, so they are provided with a light-colored screw to be used for the grounded white wire. The black and red wires in the 3-wire cable are attached to the dark-colored terminal screws, and the switch must be grounded to its box with a green or bare grounding wire.

At the second switch box, the white wire in the 3-wire cable attaches to the light-colored terminal on the 3-way switch. The red wire attaches to one of the brass screws on the switch. The black wire in the 3-wire cable is spliced to the white wire from the 2-wire cable leading to the fixture. The white wire is painted black to signify that it is being used as a hot wire. The black wire from the 2-wire cable attaches to one of the dark terminals on the switch. Both cables and the switch are grounded to the box via their bared grounding wires.

The 2-wire cable enters the fixture box, where its white wire is painted black and spliced to the black fixture wire. The black wire in the 2-wire cable is spliced to the black wire from the branch circuit cable. The white fixture wire is spliced to the white wire from the branch circuit. The cable and fixture grounding wires are bonded to the fixture box.

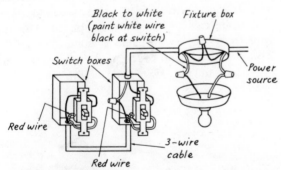

Fig. 86. Fixture controlled by different switches connected to each other (grounding wires removed for clarity).

FIXTURE BETWEEN TWO SWITCHES

This time you are trying to push 14 wires into the fixture box, so you might want to consider using a larger-size box (Fig. 87). The 3-way switches on either side of the fixture must be connected to it with 3-wire cable. Both switches are wired in the same manner, with the red and black wires attached to brass screws, and the white wire connected to the silver-colored termi-

nal. Both the cables and the switches must be connected to their switch boxes with grounding wires.

The fixture box is hooked up in this manner: The white wire from the 2-wire branch circuit cable is spliced to the white fixture wire. The black wire from the branch circuit cable is spliced to either black wire coming from the switches. The black wire from the fixture is spliced to the black wire from the opposite switch. The red wires from the two switch cables are spliced together. If you have been following all this, you might think all the white wires can be tied together, but if you do that, the switches will no longer have their own complete circuit between them. The poor little electrons rushing around all those wires will get completely confused and the whole system will not work. What you must do instead is splice together the white wires from each 3-wire cable. Don't forget to lead all four grounding wires to the box.

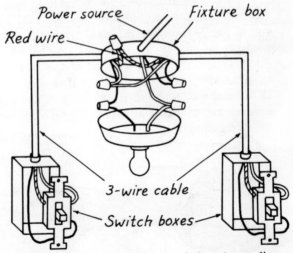

Fig. 87. Fixture between two switches (grounding wires removed for clarity).

TWO FIXTURES WITH DIFFERENT SWITCHES

You step into the kitchen and beside the door are two switches, one controlling the kitchen light and one the cellar light, both on the same circuit (Fig. 88). The switches are next to each other in the same box, but the cables between the lights, and the last light and the switches, is

3-wire cable. The first light on the 2-wire circuit cable (for example, the cellar light) has power entering it via the 2-wire circuit cable, but, again, the next link in the circuit (to the kitchen light) is 3-wire cable. The black wires in the cables are spliced together, and the black wire from the light is connected to the red wire in the 3-wire cable. All of the white wires from the fixture and cables are spliced together and all of the grounding wires are attached to the box.

The second fixture box (in the kitchen) has a 3-wire cable entering it from the cellar light and another 3-wire cable going to the switch box. The two black cable wires are hooked together. The two red wires are also spliced together. The white wire from the incoming cable (from the cellar light) is spliced to the white wire from the fixture. The black wire from the fixture is spliced to the white wire in the cable leading to the switches. The fixture and cable grounding wires are connected to the box.

The red wire from the 3-wire cable entering the switch box is attached to one terminal on one of the 3-way switches, while the white wire connects to one of the terminals on the other 3-way switch. The black wire has part of its insulation removed from its center so that it can go to first one switch and then end at the other. The switches and cable are all attached to the box via their grounding wires.

ONE FIXTURE WITH TWO SWITCHES AND AN OUTLET

The 2-wire circuit cable enters the first 3-way switch (Fig. 89). A 4-wire cable connects the first switch to the second switch and a 3-wire cable goes from the second switch to the light. From the other side of the fixture box a 2-wire cable goes to the outlet, which will always be hot. In other words, no matter whether the light is on or off, the receptacle at the end of the branch line must always be usable.

The first switch is fed by the 2-wire cable and the 4-wire cable. The two white cable wires are spliced together. The black wire in the 2-wire cable is spliced together with the black wire in the 4-wire cable and a short jumper wire which is then connected to the switch. The red and the blue wires in the 4-wire cable are attached to the switch. The switch and both cables are attached to the switch box by their bare grounding wires.

The second switch is fed by the 4-wire cable and connected to the light by a 3-wire cable. The red and blue wires in the 4-wire cable are attached to the switch. The black wires from the 4-wire and 3-wire cables are spliced together. The white cable wires are also spliced. The red wire from the 3-wire cable is connected to the 3-way switch. The switch and both cables are grounded to the switch box with their bare grounding wires.

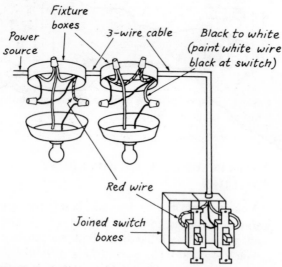

Fig. 88. Two fixtures with different switches (grounding wires removed for clarity).

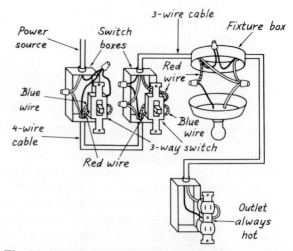

Fig. 89. One fixture with two switches and an outlet (grounding wires removed for clarity).

The 3-wire cable enters the fixture box and a 2-wire cable leaves it to connect with the receptacle. The red wire from the 3-wire cable is connected to the black wire from the fixture, while the black wires in the cables are spliced to each other. All three white wires are spliced together. The cables and the fixtures are grounded to the electrical box by bare or green wires.

The 2-wire cable from the fixture enters the receptacle box and connects directly to the outlet. The white wire is attached to the silver-colored terminal and the black wire attaches to the brass terminal. The cable and receptacle grounding wires are attached to the receptacle box.

FIXTURE CONTROLLED BY THREE SEPARATE SWITCHES

The switch nearest the light and the one farthest away from it must be 3-way switches (Fig. 90). Whether you have one or a hundred switches in between does not matter, so long as they are 4-way units. If you add more than one 4-way switch, it is wired into the circuit in the same manner as the first one.

Since the 2-wire circuit cable enters the fixture box, it also in effect connects the fixture with the nearest (and therefore 3-way) switch. In the fixture box, the black wires of the two cables are spliced together. The white wire from the circuit cable is spliced to the white fixture wire. The black wire from the fixture is spliced to the white wire of the cable leading to the first switch. The white wire is painted or taped black because it is functioning as a hot line. The two cables and the fixture are attached to the box by their bare grounding wires.

The first 3-way switch receives its power from the 2-wire cable leading from the fixture, but it is connected to the intermediate 4-way switch with a 3-wire cable. The white wire from the 2-wire cable is painted black and spliced to the black wire in the 3-wire cable. The black wire from the 2-wire cable attaches to the 3-way switch. The red and white wires from the 3-wire cable are attached to the switch. The white wire goes to the light-colored neutral terminal. The two cables and the switch are grounded to the switch box by their grounding wires.

The intermediate, 4-way switch has 3-wire cables entering and leaving it, and all 4-way switches on the circuit that you might add will be hooked up in this fashion: The black cable wires are spliced together; the red and white wires from both cables are attached to terminals on the 4-way switch; the cables and switch are grounded to the switch box with bared wires.

The last switch is a 3-way unit connected to the intermediate switch by a 3-wire cable. The white wire from the cable is attached to the light-colored terminal on the switch. The red and black wires attach to the other two "hot" terminals. The cable and switch are connected to the switch box by their grounding wires.

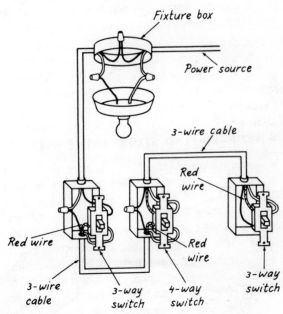

Fig. 90. Fixture controlled by three separate switches (grounding wires removed for clarity).

WIRING IN OLD CONSTRUCTION

"Old Construction" in electrical parlance is defined as the wiring in buildings that were completed *before* any electrical work was started.

Old construction is what you are working on anytime you go inside the walls of your home with intent of doing anything to your electrical system. It should be stated that the biggest problems any old construction presents are resolved more often with carpenter's tools rather than wire strippers and solderless connectors.

Whether you are functioning as a carpenter or an electrician, the first thing you have to know about any wall in your home is where to find its *studs.* If the wall is made out of plaster and lath, you can readily place your electrical boxes between studs, but it is better to secure them to something solid. If the wall is faced with wallboard, there are ways of fixing the boxes between studs, but it is preferable to nail the unit to wood. Moreover, with any kind of wall you have to know where the studs are *not,* because you can't bury an electrical box in solid wood. So when you are working in old construction as an electrician, your first chore is to be a carpenter.

WHERE TO FIND STUDS AND JOISTS

Studs are pieces of 2"×4" lumber that stand vertically in your walls. They not only support the ceiling *joists,* but both faces of the wall as well. Joists are really horizontal studs that cross over each wall and support a floor on their top edges and a ceiling nailed to their bottom edges. By common agreement throughout the building industry, joists and studs are supposed to be spaced 16" or 24" "on center" (O.C.). "On center" means the distance from the center of the narrow (2") edge of any stud or joist to the center of the next nearest stud or joist. In theory, then, if you locate a nail somewhere in the edge of a stud or joist, by measuring 16" to both its left and right you will be able to drive a nail into the next nearest framing member. Don't bet on it, however. Carpenters make mistakes too. They also get into awkward situations during the construction of a house where the studs just plain don't come out every 16" or 24". There are some general rules you can work with, though.

All doorways and windows are formed not by one stud on each side, but two, and sometimes three (Fig. 91). All of those studs may be covered by the molding around the unit, or they may extend beyond the molding. Be assured, however, that there is solid lumber on either side of any door or window, so you can begin your measuring from that point.

Another place you are sure to find 2"×4"'s is at any corner made by two walls coming together. The prescribed procedure when framing the intersection of two walls is to put three 2"×4"'s together to form a box (Fig. 92). Unfortunately, that assembly can be done in several ways, but if you go straight into a corner with your measurer the odds are in your favor that there is a stud under it and 16" away there should be another one. For certain, there will be one (and usually two) 2"×4" headers along the top of every wall where it meets the ceiling above it, and there is normally a joist above those, if the joists go in the same direction as the wall. Otherwise, the joists will cross over the header, or top plate of the wall, and be spaced 16" or 24" apart (Fig. 93).

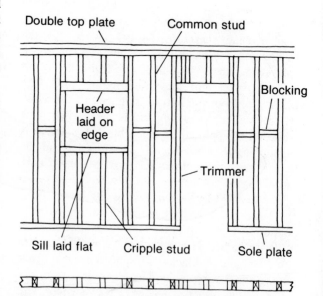

Fig. 91. Where the studs are likely to be found inside a wall.

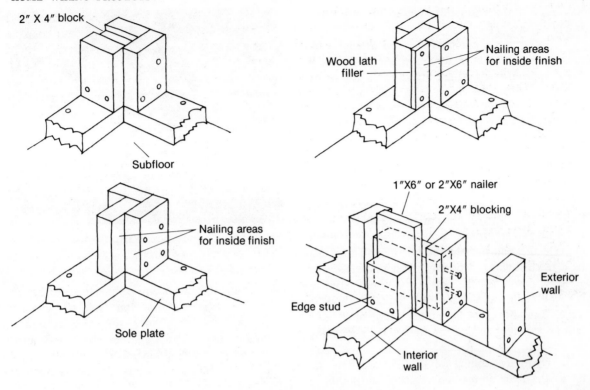

Fig. 92. Four of the ways studs are assembled at corners between walls.

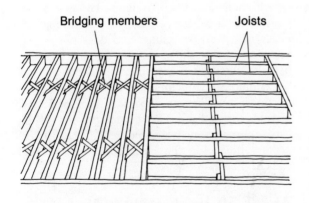

Fig. 93. Joists are more likely to be spaced exactly 16″ O.C. than studs.

SOME WAYS OF LOCATING JOISTS AND STUDS

Short of removing part or all of a wall or ceiling, you have no surefire way of locating any joist or stud. Here are some of the more reliable approaches you can take:

1. Measure 16″ along a wall from the edge of the molding around a door or window, or from a corner, and bang a 2″-long nail into the wall. You will probably hit nothing but air. Pull the nail out of the wall and continue nailing it into the wall at 1″ intervals in *both* directions until you strike wood. Then measure 16″ in the direction you want to go and start nailing again. If you find the next stud at exactly 16″, you can take a flyer and measure to the exact place where you want to put your electrical box. If the studs are 16″ O.C., there will be one at 16″, 32″, 48″, 64″, and so on. (You can fill your exploratory nail holes with daubs of joint compound, plaster of paris, or spackle.)

2. Rather than do any measuring before you make holes in a wall, you could tap along it with a hammer or your fist. When you hit a wall, it makes a hollow sound until you reach a framing member, at which point the sound becomes a dull thud. Then you can start driving nails at that point. But sound is deceptive, so don't be disgruntled if you still wind up putting a dozen

nail holes in the wall before you locate any framing members.

3. There are a number of "stud finder" gadgets on the market that you can try. Some of them are magnets that are attracted to the metal in nails holding the studs. Others require a hole drilled into the wall. Some of these devices probably work reasonably well, but they represent a two- or three-dollar expense and there is no guarantee they will perform any better than measuring and nailing does.

4. This is the one surefire way of locating studs, but use it as a last resort: Remove the baseboard along the bottom of the wall. You'll have to pry it out carefully so that you don't damage the wall or split the board, and that can be a ticklish chore. However, with the baseboard gone, you will see the bottoms of every stud in the wall at the point where it is nailed to the *sole plate*. The sole plate is another piece of 2″×4″ laid on the floor so that the studs have something to attach to.

5. If you are looking for joists in a ceiling, go up to the floor above and examine the joints in the flooring. Floorboards are usually nailed at right angles across the joists, so with some careful measuring from the wall you should be able to figure out where you can cut into the ceiling.

6. If your wall is made of wallboard, hold a light bulb close to the surface and press your ear (either ear will do) against the wall and sight along its surface. Wallboard is always nailed to the studs and sometimes by looking very carefully you will see indentations where the nails have been covered by joint compound. If you find such an indentation, move the light (and your head) up and down the wall. There should be a line of nails from the ceiling to the baseboard, spaced every 8″ to 12″. You can do this same sort of scrutinizing along your ceiling, providing you have a stepladder that is high enough to get your head up to it.

FISHING CABLE THROUGH WALLS AND CEILINGS

Probably the most exasperating, time-consuming, thankless and, if you hire an electrician to do it, most expensive part of electrical work in old construction is getting the branch circuit ca-

Fig. 94. An electrical box, when it is used as a junction for wiring, must have a metal cover to protect the wires and splices.

bles through the walls and ceilings of your house. Aside from the fact that you can't see where a given cable is going once you have poured it down a hole in the wall, the studs and joists often have bridging members between them which the cables must go through or around in some manner.

Every electrical box in your home is, or must be, connected to some other box via at least one circuit cable. In every instance, the cable leaves a given electrical box, snakes its way through the inside of a wall or ceiling, and does not emerge into view again until it reaches another electrical box. It cannot have any splices in it but must be a continuous conductor. If you do have to splice it, there must be an electrical box positioned around the splice. The procedure for fishing cable through your walls and ceilings begins with making holes for the electrical boxes at both ends of the cable run.

WHERE TO MAKE THE HOLES

You can install an electrical box anywhere in a plaster and lath or wallboard wall so long as it is not on top of a stud. The preference is always next to a stud, so the box can be secured to solid wood, but that is only a preference.

Switches are generally installed 48″ above the floor. If they are next to a door, be sure they are placed on the lock side of the door, rather than the hinged side, where they would be inconvenient to reach when entering the darkened room.

Receptacles are normally placed between 12″ and 18″ above the floor.

Wall-mounted light fixtures should be between 60″ and 70″ above the floor.

Ceiling fixtures must be placed between joists.

Always use 2½″-deep electrical boxes, rather than the shallow ones, to give yourself plenty of space for cable wires and devices.

MAKING HOLES IN PLASTER AND LATH

The plaster walls found in older homes (built prior to World War II) consist of a grid of ½″×1″ strips of softwood nailed across the studs (or joists if it is a ceiling) with a ¼″ space between strips. The strips are called *lath,* and their purpose is to provide a base for the plaster spread over the framing. The plaster is pushed against the face of the lath and into the spaces between the strips, where it hooks behind the wood. The lath is strong enough to hold an electrical box, so if you are not attaching the box to a stud, try to get about 5″ or 6″ away from the stud. Then take the following steps:

1. Outline the box on the wall. Just hold the box against the wall and trace around it with a pencil.

2. Chip away the plaster in the middle of your outline, until you can see the lath (Fig. 95). A chisel can be useful, but any gouging tool will

do, including a utility knife. Try not to break any plaster outside your outline, but don't get upset if you do. You can patch it later with spackle, plaster of paris, or joint compound.

3. When you have exposed the three pieces of lath inside your outline, place the box against the wood and trace it on the wood. The box should be centered over the middle lath strip and overlap the bottom half of the top lath and the top half of the bottom lath.

4. Remove the center lath by sawing through it with a compass or utility saw. Then cut notches in the top and bottom lath strips so the box will fit into your hole. Do *not* remove all of the top and bottom lath; you need the wood to hold the screws that will secure the box to the wall.

5. Eventually, when you have drawn a cable to the hole and connected it to the electrical box, the box should fit into the hole in the lath and be secured to the top and bottom strips with screws (Fig. 97). At that point, you can cover any bared lath around the box with plaster of paris, spackle, or joint compound.

CUTTING HOLES IN DRYWALL

Drywall, Sheetrock, plasterboard, wallboard, and a number of trade names are applied to the same material. The material consists of plaster sandwiched between pieces of heavy paper to

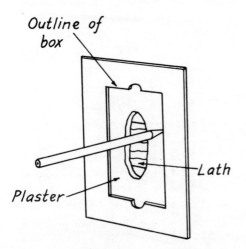

Fig. 95. Trace the electrical box on the wall and remove the plaster inside your outline.

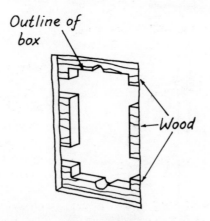

Fig. 96. Remove all of the center lath, but leave as much of the top and bottom strips as you can.

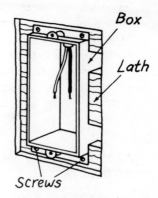

Fig. 97. Place the box in your hole and drive screws through the top and bottom lath.

form 4'×8' panels that are then nailed to the studs in a wall or the joists in a ceiling. The drywall can be anything from ¼" to ⅝" thick, so you could easily kick holes in it for your electrical boxes with your hobnailed boots, but the holes would be somewhat uneven, so use a compass saw instead.

The strongest part of a drywall construction is at its studs, and you can buy electrical boxes that have a flange attached to one side that can be nailed to a stud face.

Use the following procedure for cutting your holes:

1. Trace the box on the drywall next to a stud (Fig. 98).

2. Using a compass saw, punch a hole in the drywall and cut out the hole.

3. Push the box into the hole with its flange overlapping the stud. With the flange against the drywall, trace its outline on the wall, then remove the box.

4. Use a utility knife to cut away the drywall inside the outline of the flange.

5. When you have run cable to the electrical box, the box is placed in its hole and the bracket is screwed to the stud. You can cover the bracket with spackle or joint compound.

If you cannot place an electrical box next to a stud, you can mount it anywhere in the drywall using metal grips to hold it in place. The grips are merely strips of light metal that have long tabs on them. You place the box in its hole and slide the grips into the hole on either side of the box. Then fold the tabs over the rim of the box

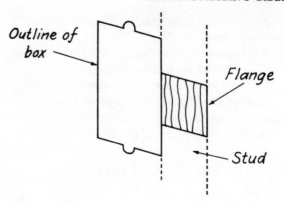

Fig. 98. Position the box so that its flange will be over a stud.

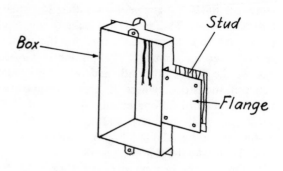

Fig. 99. Nail or screw the box to the stud, then cover the flange with joint compound or spackle.

to hold it. Alternatively, some boxes are sold with clamps attached to their sides which hold the box in place by tightening a screw (Fig. 100).

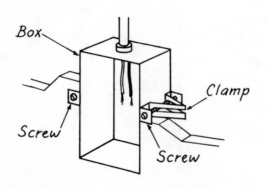

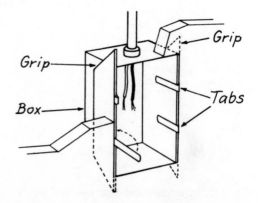

Fig. 100. Two systems for holding electrical box in wallboard when there are no framing members to nail into.

MOUNTING CEILING BOXES

Electrical boxes in ceilings present a slightly different problem in that they usually must support the weight of a light fixture, as well as the cables entering them. So you do not rely on the strength of the lath or drywall covering the ceiling to hold the box.

1. Locate where you want to position the electrical box, making certain your hole will be somewhere between joists.

2. Trace the box on the ceiling and cut out the hole (Fig. 101).

3. The box is hung from a *spanner* that is fastened across the tops of the laths or joists. The spanner is a metal strip with a bolt that can be slid along its length to position it over the center of the hole you have just made. Stick the spanner up through the hole and rest it on top of the laths or joists, then slide the bolt so that it hangs over the hold (Fig. 102).

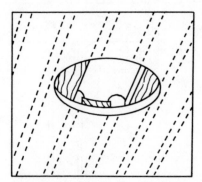

Fig. 101. Cut a hole in the ceiling laths between joists.

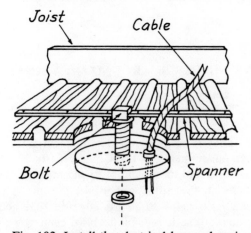

Fig. 102. Install the electrical box and position the spanner across the top of the laths on either side of it.

4. When all the cables have been run to the ceiling box and clamped in it, push the box up into its hole so that the bolt on the spanner enters the knockout in the back of the box. Then tighten a locknut to the bolt.

RUNNING CABLE

Now comes the fun part. Having made holes in your walls and ceilings, you have to run cable between each of them. The process of drilling holes through the framing members inside your walls and ceilings is the same, whether you are pulling a cable from one floor to the next, or trying to go from a ceiling fixture to a switch position in an adjacent wall.

To get from the middle of a ceiling to 4' above the floor in an adjacent wall, you may have to change tactics somewhat, but in general the procedure goes like this:

1. Go up to the floor above the ceiling fixture hole. If that puts you in an attic that has no floor your life is about to become eminently simpler, because you can just lay your cables between the joists. But assuming there is a second-story room complete with floor on top of the ceiling fixture, go to the wall directly above the wall that will have the switch box. Remove the baseboard.

2. At a position more or less above the switch-box hole, drill diagonally down through the sole plate of the second-story wall and the two top plates of the wall beneath it (Fig. 103). You will need an 18" extension bit on your drill, and you want to make a 1"-diameter hole.

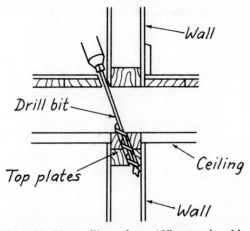

Fig. 103. You will need an 18" extension bit on your drill to get through the bottom of one wall and the top plates of the one below it.

3. Feed a 12' *fish tape* down through the hole. Then go downstairs and pull the end of the tape out the switch-box hole. The tape is a flat strip of steel about half an inch wide that is relatively easy to push through the spaces between studs and joists.

4. If the joists run at right angles to the walls, feed a second tape through the ceiling-fixture hole to the wall and hook it around the first tape. Then pull on both tapes until, theoretically, their ends hook together.

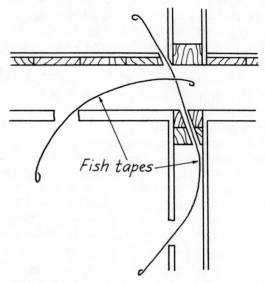

Fig. 104. Work your fish tapes toward each other and hook them.

If the joists run parallel to the walls and there is no subfloor, remove one floorboard that is between the joists on either side of the ceiling fixture hole. Hopefully, one board will give you enough space to drill 1" holes through each joist so you can thread a cable from the fixture to the wall. If drilling is not possible, cut a ¾"×¾" notch in the top of each joist. If there is a subfloor, you probably cannot get to the joists, so don't bother.

5. Once the two fish tapes have hooked together, hook the wires in a cable around the bent end of the fish tape coming out of the ceiling box, and pull the tapes and cable through the fixture hole, down the inside of the wall, and out the switch-box hole.

If you have had to drill or notch the joists, the cable need only be pulled up through the wall

and then threaded through the holes in the joists.

One variation of the above procedure occurs when you have no access to the floor above the ceiling, or if there is a subfloor under the floor and you cannot get at the joists. In this situation, cut a large hole in the wall about 6" from the ceiling above the switch hole. Drill straight up through the top plates of the wall and into the cavity between the joists. Then run a fish tape through the hole and try to hook it on to a tape coming across the ceiling from the fixture hole. If you find the joists are running across your path, and you cannot get at them by removing part of the floor above, you will have to cut a channel in the ceiling from the fixture hole to the wall and then either notch or drill through the joists.

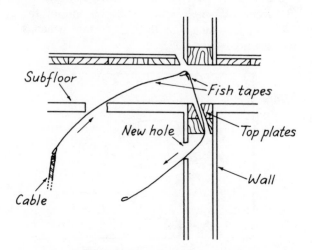

Fig. 105. Tie the cable to one of the tapes and draw the tapes (and the wire) through the ceiling and walls.

RUNNING CABLE AROUND OBSTACLES

The studs in walls do not all just stand solemnly by themselves with 16" of space between them waiting to be filled with electrical cable. They often have bridges made of 2"×4" lumber nailed between them, particularly if the house is old and the weight of its plaster and lath might twist the studs out of alignment. The bridging also helps to retard any fire that gets into the wall, so it is liable to have been put almost anywhere between any two studs.

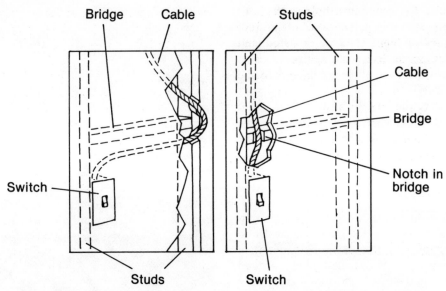

Fig. 106. Two ways to cut paths for electrical cable around framing members.

If you drop a cable down the inside of a wall and it comes to a halt a third of the way down, you have encountered a piece of bridging. Try to locate it precisely by marking the cable, pulling it out of the wall, and measuring the distance from your mark to its end. Now go down to the point in the wall where your measurements indicate the bridging is located and cut a hole in the plaster covering the bridge. When you have exposed the bridge, notch it by cutting a ¾" square hole in it. Lower the cable, feed it through the notch, and keep pulling it down to wherever it is to emerge from the wall (Fig. 106). You might, by the way, run into two or three bridges in any given wall before you get from the ceiling to an outlet box near the floor.

RUNNING CABLE ALONG A WALL

When you have to run an electrical cable from one end of a wall to the other, the cable must travel past a stud just about every 16" or so. In new construction work, the studs have not been covered until an electrician has had a chance to make his installations, so all the electrician needs to do is drill 1" holes through the studs and run his cable. But when the studs have been covered with plaster and lath or wallboard, the task becomes a little more time-consuming. Now what you have to do is make a channel in the face of the wall around the front of each stud. With a plaster and lath wall you may find the easiest approach is to rout a trench in the plaster, lay the wires inside the trench, and cover it with plaster (Fig. 107).

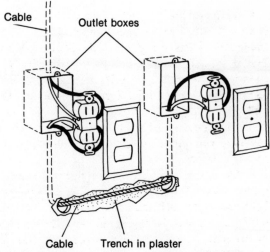

Fig. 107. You can gouge a trough out of plaster to get from box to box, bury your cable, and cover it with plaster.

Alternatively, you can drill a hole in the wall face and draw the cable as far as the next stud, then drill holes on both sides of the stud and notch the stud so that you can bring the cable out of the wall, past the stud, and back into the wall again. This is also the way of dealing with drywall construction. With a little luck, you might be able to remove the baseboard at the bottom of a wall and run the cable against the bases of the studs, then replace the baseboard.

GETTING AROUND DOORFRAMES

When you need to carry a circuit cable from one side of a doorway to the other, especially in older homes, you may find the work easier than you would imagine. Remove the molding around the doorjambs and header. Normally there is a considerable space between the framing members under the molding that allows you to run your cable up one side of the doorway, across its top, and back down to the other side, so all you have to do is drill holes through the studs to get the cable in and out of the channel (Fig. 108).

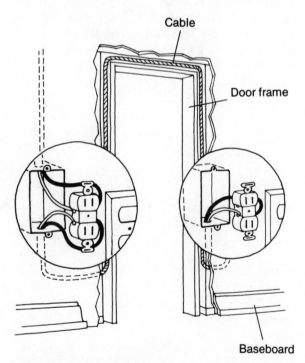

Fig. 108. There is often a space between the frame and jambs of doors that is wide enough to hide your cables.

SURFACE WIRING

By far the easiest way of adding electrical service to any part of your house, as well as the least expensive, is to install surface wiring. The equipment you need for this is not very attractive sitting on the surface of the wall, but it is not un-

Fig. 109. Although this ceiling box and fixture are mounted in full view, they could have been hidden in a ceiling.

sightly, either, and in many instances you can be relatively discreet about where you place it so that it does not become visually obtrusive.

PLUG-IN STRIPS

Plastic strips are plugged in to any wall outlet and then fastened along a baseboard with screws. At whatever point you need an outlet, you can insert a molded plastic receptacle.

MULTIOUTLET STRIPS

Another version of the plug-in strip, multioutlet strips can be fastened to a wall or baseboard with screws and are ideal over home shop workbenches or kitchen counters. The strips are a flexible plastic ribbon containing circuit wires which are connected directly to a branch circuit cable at an outlet box. The black wire from the multioutlet strip connects to the black cable wire, and the two white wires are also spliced together. The ribbon is designed with parallel slots running its full length, and wherever you need an outlet, you simply twist a plastic outlet into place between the slots.

Fig. 110. Multioutlet strips can be mounted on masonry walls as well as a baseboard. The receptacle can be twisted out of its track and placed anywhere along the strip.

METAL RACEWAYS

There are all kinds of metal raceway designs. Some are small square conduits, others are metal or plastic strips that are separable. Metal raceways have been around for a long time and are widely used to electrify outbuildings on farms, where attractiveness is less important than utility. The raceways can be fastened to the surface of any wall, floor, or ceiling; convenience outlets and switch and fixture boxes are mounted to the channels and affixed to the wall surface. The convenience boxes manufactured for surface mounting do not have any sharp corners the way regular electrical boxes have, but otherwise they look about the same.

Normally, metal raceways are connected directly to a branch circuit cable and wire is then pulled through them. Or, if they are the type that comes in two sections, the wires are laid in the bottom half and the top half is then snapped on over the wires. In any event, metal raceways are extremely convenient and easy to use.

INSTALLING RECESSED FIXTURES

Recessed ceiling fixtures (such as fluorescent lights) have a variety of decorative advantages, but they must have space around them for air to circulate so that heat from the bulb can dissipate and not become so intense it causes a fire. Installing a recessed fixture is a relatively straightforward procedure:

1. Locate where you want the fixture, but be sure the position is between joists before you cut any holes in the ceiling.

2. Trace or mark out the dimensions of the hole needed for the box, then cut out the hole with a utility or compass saw (Fig. 111).

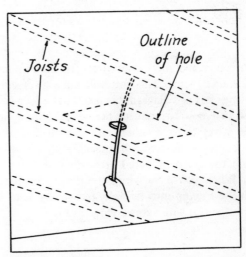

Fig. 111. Cut a hole in the ceiling for your recessed fixture box.

3. Some manufacturers provide metal strips to support the fixture box between joists. If the unit you are installing has them, install them. Otherwise screw 1″×2″ wood strips between the joists, positioning them so they form a wooden frame in the ceiling that is exactly the size of the box (Fig. 112).

4. A junction box may or may not have been provided by the fixture manufacturer. If one did not come with the unit, buy it separately. The cable leading into the junction box is connected with a cable clamp; leave at least 6″ of wire inside the box.

5. The junction box is normally united to the

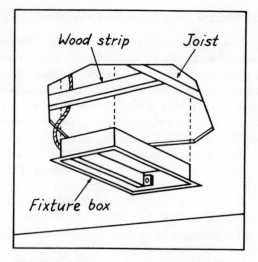

Fig. 112. Frame the hole with 1"×2" or larger strips of wood.

fixture box and the cable wires are then spliced to the light wires. The unit will have an accompanying wiring diagram, which in effect will show you how to connect the white fixture and cable wires to each other, and the black fixture and cable wires to each other, and where to attach the green or bared grounding wires.

6. When you have wired the fixture properly, push the box up into its hole and secure it to its support straps with screws (Fig. 113).

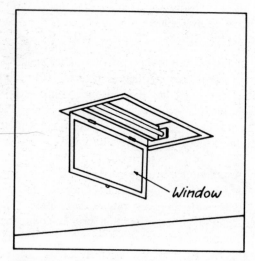

Fig. 113. Screw the fixture box into the frame.

INSTALLING A FLOOR OUTLET

Floor outlets are a marvelous convenience for equipment such as stationary power tools in the middle of large rooms, and as a place to plug in work islands in kitchens. They are unusually easy to install, particularly anywhere in a ground floor that is directly over a basement where the branch circuit cables are running along the joists and are easily reached.

1. Locate where you want to place the floor receptacle, then go downstairs with a ruler to make sure the box position will not be over a joist. Drill holes at each corner of the rectangle from the basement, so you will be absolutely certain to miss any of the joists.

2. Cut out the box with a compass saw (Fig. 114).

3. You have to take your power from one of the branch circuits. One way to do this is to *turn off the power* in one of the cables and then sever it, bringing both ends into the floor box to hook them to your floor outlet. However, you may not have enough cable to accomplish that, so your next possibility is to sever a cable and reattach the ends in a junction box along with a separate piece of cable that is long enough to reach into the floor receptacle box. In the junction box, all of the white wires are spliced together and all of the black wires are spliced to each other. All of the cables are grounded to the box with their bare wires.

4. Run the cable or cables up through the hole in the floor and clamp them into the floor box, then screw the box to the floor (Fig. 115).

5. Floor receptacles are recessed in their boxes to allow for a dust cover that screws through the faceplate to protect the outlet when it is not in use. Otherwise, the outlet is identical with any single outlet receptacle. The black wires are attached to the brass terminal screws and the white cable wires go to the silver-colored terminals. The cables and outlet must also be grounded to the box with bared grounding wires.

6. Install the floor outlet plate over the receptacle, screwing it in place to the electrical box (Fig. 116).

Always use appliance plugs that are narrow enough to fit well inside the receptacle box and can be protected by a dust protector inserted around their cords (Fig. 117).

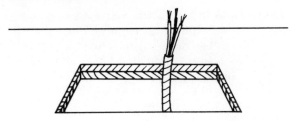

Fig. 114. Bring the cable up through the hole in your floor.

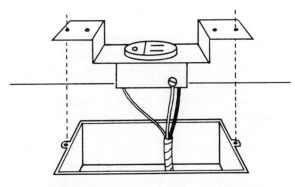

Fig. 115. Install the electrical box and the floor receptacle.

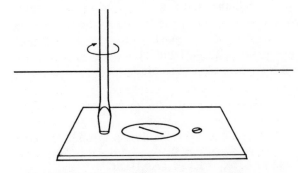

Fig. 116. Use a specially designed floorplate to cover the outlet.

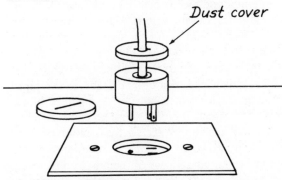

Dust cover

Fig. 117. Use appliance plugs that are thin enough so that you can use a dust cover over it when it is plugged into the outlet.

OUTDOOR INSTALLATIONS

Outdoor receptacles, switches, and fixtures must be weatherproof. What makes the switches and outlets that way is usually a spring-loaded cover that seals the electrical box, as well as calking around the juncture between the coverplate and whatever surface it touches. In many communities you are also required to protect all outdoor cables in weatherproof conduit, and surface-mounted equipment must be installed in a weatherproof electrical box that has all of its unused knockouts closed with threaded plugs or spring-loaded covers.

All NEC grounding and wiring rules for indoor wiring apply outdoors. The only difference is the equipment and materials that you use; you can only use fixtures that have been UL-listed for use outdoors, for example. Even the bulbs you put in the fixtures must be weatherproof types, so they will not shatter the first time they are struck by rain or snow. For that matter, any tools or appliances you plug in to an outdoor receptacle should be either double-insulated or have a 3-prong grounding plug. And, if there is no ground-fault circuit interrupter controlling the entire outdoor circuit, there should be one plugged into the outdoor outlet.

Fig. 118. Part of the weatherproofing of an outdoor receptacle is a spring-loaded cover which protects the outlet when it is not in use.

BRINGING POWER OUTDOORS

There are several ways that you can draw power to an outdoor installation, although the preferred method is to give your exterior installations their own circuit that begins at the distribution panel or an add-on distribution box with a GFCI. The outdoor circuit cable is run from the distribution panel through the outside wall of your house (Fig. 119).

Here is the basic procedure:

1. The circuit cable from the distribution panel ends at a grounded junction box attached to the interior of an outside wall of the house.

2. The indoor junction box is connected to a junction box or receptacle on the outside of the house via a galvanized conduit pipe. The pipe is threaded into the center hole in the back of each box.

3. The outdoor cable enters the outdoor box, passes through the linking piece of conduit to the indoor junction box, and is spliced to the indoor cable.

If you are unable to provide your outdoor electrical system with its own position in a distribution panel, you can take power from an indoor branch circuit nearest where you want to go outside. You will still need a junction box and it still ought to be connected to a box on the outside of the house with a short piece of conduit. And you still must use a GFCI.

INSTALLING AN OUTDOOR OUTLET

1. Wherever you take the power from, bring it to a junction box indoors at the point you want to exit the house. You might include a switch in the box to control the outdoor outlet.

2. Install the junction box and cut a 1″ hole through the outside wall of the house for the conduit.

3. Cut a hole exactly the size of the outdoor outlet in the exterior wall of the house.

4. Install the outdoor electrical box and connect a short length of conduit between the two boxes, securing it to the boxes with locknuts.

5. Install the outlet in the outdoor box, and connect it to the switch in the indoor box.

6. Cover the receptacle with a weatherproof coverplate. Calk around the plate to prevent water's getting at the receptacle.

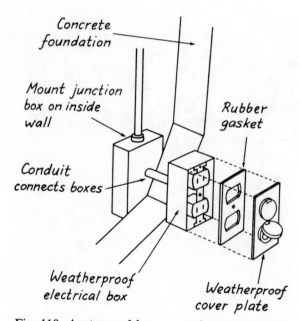

Fig. 119. Anatomy of how an outdoor receptacle is connected to the house wiring system.

INSTALLING FREESTANDING ELECTRICAL BOXES

Freestanding receptacles can be useful in or near a patio or at various places around your yard. They should stand at least 18″ above the ground, and one way to keep them erect is to attach the electrical box to the threaded end of a vertical galvanized conduit. To stabilize the pipe, imbed it in the center hole of a concrete block filled with cement (Fig. 120). The cable running to the box can be protected by conduit or not, depending on your local code requirements. It should be buried at least 6″ below ground level, and preferably 18″, to protect it against damage. The concrete block is, of course, also buried in the ground.

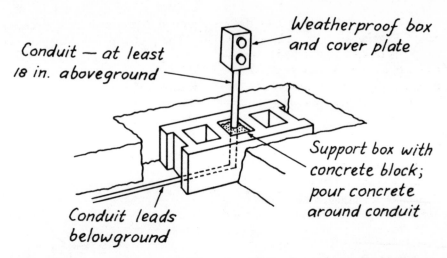

Conduit — at least 18 in. aboveground

Weatherproof box and cover plate

Support box with concrete block; pour concrete around conduit

Conduit leads belowground

Fig. 120. Stand outdoor receptacles in the center of a concrete block. The conduit that holds the receptacle is buried in cement.

INSTALLING EAVE LIGHTS

Eave lights, even though they reside under the overhang of your roof and are somewhat protected from the weather, must still be installed in weatherproof fixtures. You can most likely bring power to the eave light from a junction box in your attic and connect the circuit cable to the fixture by running it through a 1″ hole drilled through the eave. Clamp the wire to the fixture box and secure the box to the underside of the roof overhang.

You can control the light by installing a photoelectric eye adapter, which is threaded into the socket of the fixture. A weatherproof light bulb is then screwed into the socket in the photoelectric eye device. The eye will automatically turn the light on at dusk and off at dawn.

You can also run a cable from the attic junction box downstairs to an indoor switch that controls the light.

Fig. 121. Outdoor lamps must be capable of withstanding the elements.

INSTALLING A HEAT TAPE

Heat tapes are attached along the bottom edge of a roof and laid in the house gutter system to prevent snow and water from freezing and building up ice dams along the eaves, where the ice can work its way under the roofing and drip inside the house. The heat tape is electrically controlled, which means it has to be plugged into a receptacle. The most convenient place for such a receptacle is under the eaves. You can install the

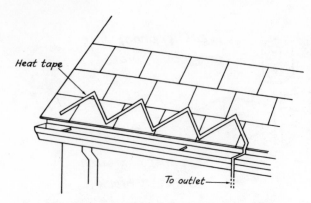

Fig. 122. Heat tapes are installed along the gutter and eaves, and must be plugged into an outdoor outlet.

receptacle in the same manner you would an eave light, by picking up power from a junction box in the attic and carrying the cable through a 1″ hole in the eaves. Because heat tapes must be turned on and off manually, you will also need to wire the receptacle into a conveniently located switch.

INSTALLING A POST-MOUNTED LAMP

A post lamp stationed away from your house, say at the end of the driveway, must receive its current via an underground cable, and there should be a switch inside the house that controls the lamp. The switch actually should be near the point where the outdoor cable exits the house.

When you are burying your conduit or weath-erproof cable, it should be about 18″ below-ground to protect it from gardening tools.

The post that holds the outdoor lamp must be sunk in a hole (Fig. 123). Line the bottom of the hole with 3″ of bricks or stones, then fill it with concrete and stand the lamppost upright in the concrete. The post, of course, is hollow to allow the cable wire to reach the light socket. The cable must be grounded and should be protected by a ground-fault circuit interrupter.

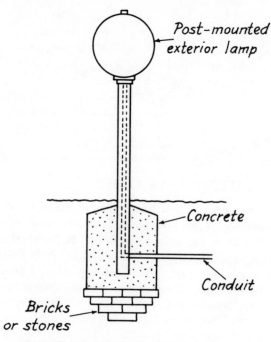

Fig. 123. Anatomy of an outdoor post lamp.

CHAPTER SIX

Electrical Repairs Around the House

Repairs around the home involve working primarily with lamps and signaling systems. Lamps, even the ones that are permanently tied into your house system, have their own self-contained wiring circuits. In most cases, the lamp can be removed from its power source merely by pulling a plug out of a wall receptacle. Signaling systems include bells, buzzers, and chimes, and while they receive their power directly from the house system, their circuits are very low voltage arrangements using ⚡16 or ⚡18 bell wire.

HANGING CEILING FIXTURES

One of the most common of all electrical chores is replacing a ceiling fixture. Opening up a forty-year-old fixture box and pulling out a maze of black, dusty cable wires can be one of the more perplexing experiences in life. All of the wires look alike, and if there is a wall switch or two controlling the unit, and/or another branch circuit going through the box, your first reaction is to forget the whole project. To make matters worse, when you look at the new fixture you want to install, you discover it has more than two conductors and you have no idea where the extra three or four wires are supposed to be connected.

In the first place, remember that with the exception of wires coming from a switch, all white wires go together, and all the black wires are connected to each other. If the new fixture has more than one bulb in it, there is likely to be a separate set of white and black wires coming from each lamp, but when they meet the circuit

Fig. 124. Hanging ceiling fixtures can be practically any design imaginable.

cable, they all have to come together. Finally, no matter what kind of morass you find in the electrical box, the fixture presently there is (or has been) working; it is correctly wired to the various cables, and never mind how black all of the cable wires appear to be. Moreover, if you wire your new fixture incorrectly the worst that can happen will be that it blows a fuse or doesn't turn on at all, in which case you know the wires are incorrectly hooked and you simply have to rearrange them.

If the new fixture has its own switch and there is no wall switch involved, all you need to do is locate the white wires in the electrical box and splice them to the wires from the new fixture. Similarly, splice the black fixture wire(s) to the black wires in the electrical box. If you cannot tell by looking at the wires in the box which are white and which are black, you should be able to figure out their color by where the wires in the old fixture are attached. If there is an extra green (or green and yellow) wire in your new fixture, it is a grounding wire and is attached to the electrical box.

If the new fixture is to be controlled by a wall switch, you have another set of wires coming into the electrical box to worry about. The black wire(s) from the fixture must be spliced to one of the two wires coming from the switch. It does not really matter which switch wire you connect because both are hot wires, even though one of them is white and may or may not have been painted black or covered with black tape. The other wire from the switch is spliced to all of the black cable wires in the box, but if you are only replacing a fixture you don't care what it does because you don't even have to find it, let alone do anything with it. All of the white wires (except the painted one coming from the wall switch) are spliced together, including the white wire(s) from your new fixture. In other words, all you have to do is remove two wire nuts (or tape). One of them holds one black wire from the switch and another from the fixture, and the other contains all of the white return wires.

While the wiring in a fixture can be confusing, the real labor in installing a new overhead fixture comes from working with your arms over your head. Your arms will get tired. You have to wrestle a myriad of wires and then hoist your new fixture up in the air and keep it there long enough with one hand to make the wiring connections using your other two hands. Furthermore, the way in which the new fixture is held to the electrical box is, for some mysterious reason, nearly always different from the way the old fixture was held there. The solution to all these dilemmas is to plan ahead—way ahead.

Make sure that all of the power coming into the fixture box has been turned off. Then undo the old fixture from its moorings and pull it down from the box far enough so that the cable wires it is attached to come out of the box. Now just sit there on top of your wobbly stepladder and take a good look at how the old fixture was held to the electrical box (Fig. 125). There might be a steel bracket that spans the box and is held in place by a threaded bolt emerging from the center hole in the back of the box. The bolt can just be unscrewed if you want to get rid of that arrangement. The bracket might be attached to the sides of the box by screws. There might not be any bracket at all, but a screw nipple threaded on the bolt that holds a chain leading down to the fixture.

When you have figured out how the old fixture was installed, look at your new unit and determine how the manufacturer wants you to hang it. You should find an assembly diagram (and a

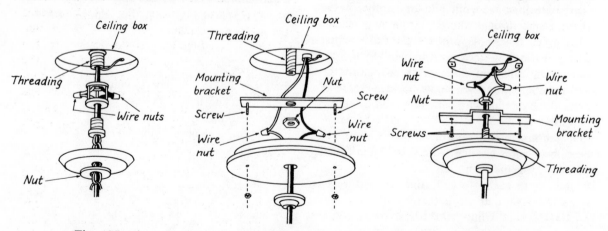

Fig. 125. Three of the many ways that ceiling fixtures are held over their electrical boxes.

wiring scheme) somewhere in or on the box the fixture came in. Consult the diagram; it may be very enlightening. It may also be more confusing than helpful. Whatever you do, plan exactly how you intend to hang the new fixture before you do any work. If the fixture wires must travel up a chain or a pipe, determine how they emerge in the electrical box. They may, for example, have to come through the bottom part of a nipple and then out a hole in its side. Or they may have to go around a bracket. Whatever must happen to them, know what you have to do before you start hanging the unit. Also determine which wires are connected to what cable wires. If you are hesitant about the wiring, plan on removing one wire at a time from the old fixture and replacing it with a corresponding wire from the new fixture. Nevertheless, there will still come a moment in your life when you need to hold up both fixtures and work on the wires simultaneously, and you conceivably could run out of hands.

When you know exactly how you are going to proceed during the process of changing fixtures, you can begin working. One of the ways you can make your life considerably easier is to tie a wire or strong cord to the electrical box and use it to support the new fixture while you are working on the wiring. Or you might be able to rest the unit on top of your stepladder, or have an assistant hold it. By and large, you want to have your hands free to work with the wires and screws in the box, so figure out some way of supporting the fixture. Also prepare all the wires you will need to handle. Twist the white ones together and the black ones together and straighten them out or thread them through the hanging chain and nipple, or spanner, or whatever, so that when you raise the fixture up to the box all you need to do is wrap them around the wires in their proper splice and secure them with wire nuts.

You have to be inventive and a little bit crazy to hang a new ceiling fixture, especially in an old house, but it can be done with a lot of patience and some good planning.

ELECTRIC CORDS AND PLUGS

Most lamps and appliances receive the electricity necessary to operate them via a plug and a 2-wire power cord. The plug has two or three metal prongs which make contact with the electrical power flowing through a receptacle in the wall of your house. The electricity flows into the electrical device through one wire in the cord, follows the electrical circuit in the device, then returns to the house electrical system through the other wire in the cord.

Power cords and their plugs are vulnerable to abuse. People jam them into wall outlets, then yank on the cord to get them out again. The cords are stepped on, wrapped around things, wiggled, pulled and bent endlessly. Often, the insulation around the cord wires wears away, exposing the wires; or the wires, even if they have been molded to the plug with plastic sheathing, are pulled loose of the plug. Whenever you have a lamp or appliance that causes a short circuit, or the unit just plain won't work, look first to the cord and the plug. The odds are better than fifty–fifty that the trouble is an exposed or broken wire, or a loose connection between the wire and either the appliance or the plug.

SPLICING CORD WIRE

Cord wire is normally stranded so that it will have considerable flexibility. There are usually two wires in a power cord. Often they are uncolored, and are merely bare stranded wires twisted together and covered with plastic. Some power cords are manufactured to carry heavy voltage loads and the wires in these have their own insulation inside a sheathing. But the insulation usually is all the same color (black) since it does not matter which one is used as the hot line and which one functions as the return. In the case of some modern appliances where the grounded side of the unit's circuit must be connected to the white-wire side of the outlet, the plug attached to the cord has different-sized prongs to correspond with the different-width slots in a modern 3-hole outlet.

When a line cord is frayed and one of its wires is visible, it can often be protected by wrapping the cord with plastic electrician's tape. If both wires are exposed, there is a good chance one bare metal may touch the other, causing a short circuit. In this situation, the wires should be wrapped separately, and then together. Some types of power cord, known collectively as zip cord, are designed with insulation around each of the wires, but the insulation is attached by a thin membrane of plastic. To separate the wires, you only need to pull them apart. This allows you to wrap tape around each wire, then tape the wires together again. If the cord you are repairing is not easily divisible, the easiest way of repairing may be to cut the cord and then splice it together again. Here's how to do it:

1. Obviously, don't try splicing a cord that is plugged into the house circuit. Remove the plug from its wall receptacle.

2. Cut the cord at its worn portion.

3. Strip the wires from the cord sheathing, so that you have about 2″ of separated wires extending from each end of the severed cord.

4. Cut one wire in each end of the cord 1½″ shorter than the other wire.

5. Strip about ½″ of insulation from the ends of all four wires.

6. Twist the bared end of one long wire with the bared end of the short wire on the second cord. With stranded wire, first twist the strands on each wire tightly in a clockwise direction. Then hook the two wires around each other at their centers and twist the wire ends around each other.

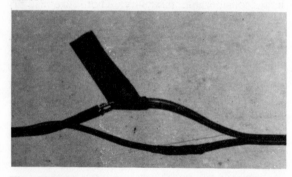

Fig. 126. Make your splices in a cord at different positions so the wires you have bared have no opportunity to ever touch.

7. Splice the remaining long and short wires together. You should now have two wire splices in the cord that are about 1½″ apart, so they in no way touch each other.

8. Wrap plastic electrician's tape around each of the splices. You will have to divide the wires enough to push the tape between them as you do your wrapping.

9. When the splices have been wrapped separately, wrap tape spirally around the cord. You are doing this partially for strength but mostly for appearance, so you only need two or three layers of tape.

Fig. 127. Wrap each splice individually, then tie them together with an outer coating of tape.

REPLACING A PLUG

Often cord wire frays or breaks at the point where it enters the plug. Some plugs are molded to their wires with the same plastic that insulates the cords, and the only way you can replace them is to cut them off the line. Some replacement plugs are quick-connect types which have a lever at their back that you lift up. The cord wires are not stripped but merely placed inside the plug and the lever is closed over them. The lever has metal teeth on it that bite through the cord insulation and make contact with the metal wires.

The quick-connect plugs are fine for most lamps and small appliances, but a more substantial (and traditional) type of plug has screw terminals. The cord wires must be pushed through the back of the plug, then wrapped around its prongs and secured to the screw terminals. But before you make the terminal connections, the wires should be tied in a strain-relief or Underwriters knot to protect the terminal connections from coming apart when someone jerks the plug out of a wall receptacle (Fig. 128). The procedure for replacing a cord plug is this:

1. Disconnect the plug from any source of electricity.

2. Remove the old plug from its cord, either by cutting the wire or loosening the terminals in the prong side of the plug.

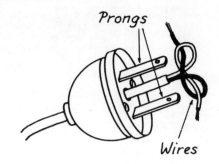

Prongs

Wires

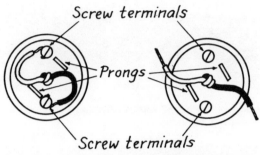

Screw terminals

Prongs

Screw terminals

Fig. 128. How to tie the Underwriters knot, and attach wires to screw terminals.

cover that slides over the prongs (it has slits in it) and covers the terminals.

Fig. 130. Pull the plug against the knot and wrap each wire around a terminal screw.

3. Remove the cord sheathing from about 2" of the wires and strip ½" of insulation from the end of each wire.

4. Push the wires through the hole in the back of the replacement plug. Tie the wires in the figure-eight Underwriters knot. The knot should be large enough so that when you pull on the cord, it cannot free itself from the plug. Pull the cord through the plug until the knot sticks in the hole and is below the surface of the plate surrounding the prongs.

5. Wrap each wire clockwise around a prong and secure it under one of the screw terminals.

6. There should be a plastic or cardboard

Fig. 131. Protect the terminals with a plastic or cardboard cover that fits over the prongs of the plug.

HOW TO REPAIR ANY LAMP

All lamps, floor or table models, antiques or contemporary, no matter what materials they are made of, or how they look, are constructed in exactly the same manner. Either they have a pipe that extends from their base to the light socket and contains the power cord, or the cord hangs down outside the lamp from the light socket. Sometimes the cords must be replaced. Occasionally the light socket becomes faulty. Most often the switch in the light socket stops functioning and the socket must be replaced.

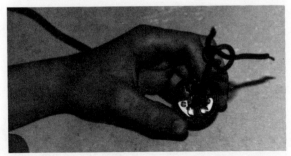

Fig. 129. Tie the wires in the figure-eight Underwriters knot.

Fig. 132. All lamps, no matter what they look like, are assembled in the same basic manner.

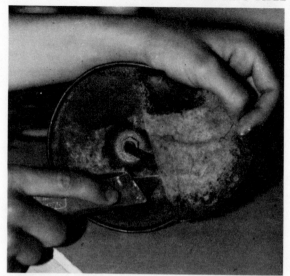

Fig. 133. Carefully remove the felt pad glued to the bottom of the lamp.

Fig. 134. Undo the locking bolt threaded to the bottom of the center pipe.

HOW TO REPAIR AN ENTIRE LAMP

1. Unplug the lamp and turn it over so you can work on the underside of its base.

2. There is normally a felt or some other type of pad that has been glued to the base. Use a sharp knife to slice between the pad and the base. Be careful not to damage the pad; you will be able to glue it back on the base after your repairs have been made.

3. When you get the pad off the base, you will see the end of the pipe that holds the whole lamp together. It is threaded and has a locknut on it. Remove the locknut.

4. All of the components of the lamp are, in effect, strung on the center pipe as if they were beads. You can slide all of the parts off the pipe, which will leave the pipe threaded to the base of the light socket.

5. Unscrew the light socket from the pipe. As you pull the socket away from the pipe, the cord will come with it as far as the plug can go before it reaches the bottom end of the pipe.

If all you are doing is replacing the light socket, you may be able to simply unscrew it from the pipe without taking the whole lamp apart. If that is the case, you can start your repairs with this step.

6. When the socket is free of the pipe, twist, then pull the brass or silver-colored shell away from its base cap. All socket shells twist into their bases and they all come apart by twisting and pulling at them.

7. Remove the cardboard sleeve that surrounds the socket proper.

8. The cord is connected to terminals on the sides of the socket. Loosen the terminals and free the cord wires.

9. Sockets cannot usually be repaired, so you replace them. You can buy a number of different types including ones with chain pulls, rotating switches, or push-button switches. If you wish to use 3-way bulbs in the lamp, you have to have a special 3-way socket. You also have a choice of two colors, brass or silver. Whatever replacement socket you are using, dismantle it first. Twist/pull off the cap, remove the insulating sleeve, and open the terminal screws.

10. If you are replacing the cord, pull the old cord out of the pipe. Then snake your new cord through the pipe. Separate about 2″ of the wires and strip ½″ of insulation from each wire.

11. Slide the socket cap down over the wires and screw it to the end of the pipe. Some socket caps have a lock screw through their bases which should be tightened against the pipe to keep the unit from rotating.

12. Connect one wire to each terminal at the bottom of the socket.

13. Slide the insulating sleeve over the socket and hold the socket against its cap. Insert the outer metal shell over the insulating sleeve and twist/push it into the socket cap.

14. String the components of the lamp on the pipe and tighten them together by tightening the locknut to the pipe at the base of the lamp.

15. If your new cord does not have a plug on it, install a plug.

16. Coat the bottom of the lamp base with ordinary white glue and stick the felt pad back on the lamp.

17. Plug in the lamp and try it. If you blow a fuse when you turn it on, you have a wire in the socket or the plug that is touching something it shouldn't and is short-circuiting the unit. Dismantle the lamp and check all four of your wire connections.

Fig. 135. All of the components that make up the lamp are strung on the pipe like beads on a necklace.

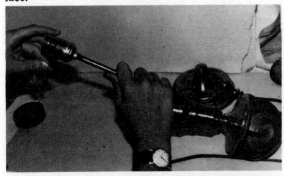
Fig. 136. Unscrew the lamp socket from the end of the pipe.

Fig. 137. The anatomy of a light socket. The outer shell fits over a cardboard sleeve that covers the socket itself. The shell twist-locks into the base cap.

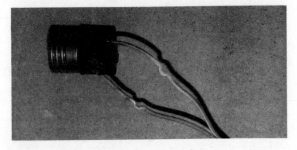
Fig. 138. The socket has two terminal screws at its base.

84

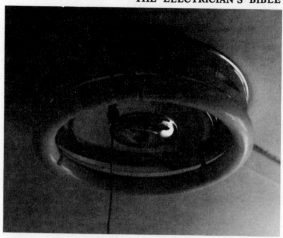

Fig. 140. Some fluorescent lamps are circular.

Fig. 139. When you have completed your repairs, plug in the light and test it.

FLUORESCENT FIXTURES

There are two basic kinds of fluorescent fixtures, starter types and rapid-start types. Both are manufactured for straight, U-shaped, or circular tubes, in a variety of sizes. The straight tubes range from a few inches all the way to 8′ in length.

ANATOMY OF FLUORESCENTS

Fluorescent fixtures consist of a glass tube (the bulb), filled with a small quantity of argon or some other highly purified gas, droplets of mercury, and coiled tungsten cathodes (Fig. 141). The bulb begins to light when the voltage between the cathodes is enough to form an arc in the gas that fills the tube. As current passes through the gas, energy is released from the mercury ions and radiates to the walls of the tube, which are coated with phosphor. This makes the phosphor give off light.

In order to generate enough voltage to ignite

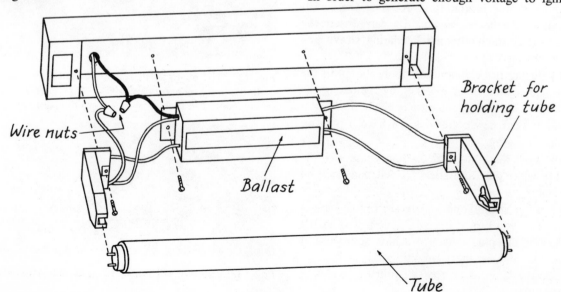

Fig. 141. Wiring hookup for a rapid-start fluorescent lamp.

Fig. 142. The starter is a small, very light canister that is fitted into a socket under the bulb.

the fluorescent bulb, the cathodes have to be preheated. The starter, which is a small, light canister, applies enough current to the cathodes for a few seconds to preheat them enough so that they will produce voltage. Then the starter removes itself from the circuit, causing the voltage to be applied between the cathodes and thereby striking an arc. To overcome the slowness of the preheat system, the rapid-start lamps light immediately, eliminating the need for a starter in the fixture.

The *ballast* found in all fluorescent lamps is fundamentally a single winding of wire around a laminated iron frame. The ballast performs two separate functions. The first is to momentarily increase the voltage coming from the house circuit to overcome any resistance between the cathodes at each end of the lamp and establish

Fig. 143. The ballasts are heavy little cubes and there will be one for each lamp in the fluorescent unit.

an arc of voltage between them. Then, when the bulb has lighted, the resistance of the path between the cathodes becomes extremely low, so the ballast acts to limit the current to a safe value. Without a ballast, fluorescent tubes would pull in so much current they would destroy themselves.

INSTALLING FLUORESCENT LAMPS

Fluorescent lamps can be used anywhere in your home to provide a broad, even lighting, such as over a kitchen work space, or in a bathroom or workshop. Fluorescent lamps will last 8 to 12 times longer than incandescent lamps, but frequent turning on and off will wear them out. As a rule of thumb, to conserve electricity and your light bulbs, turn off incandescent lamps whenever they are not needed; however, if you intend to use a fluorescent bulb again within 15 or 20 minutes, leave it on.

Connecting a fluorescent fixture to an electrical box in your house wiring system becomes a matter of splicing the black fixture wire with the

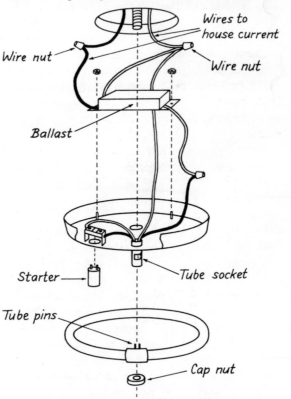

Fig. 144. Wiring hookup for a starter-type fluorescent.

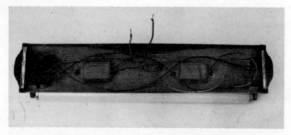

Fig. 145. Sometimes when you buy a fluorescent and take it home you discover half the wires are not connected. However, there is a wiring diagram pasted somewhere inside the unit.

black cable wire, splicing the white fixture and cable wires together, then grounding both the fixture and the cable to the fixture box. When you buy a fluorescent fixture and open it up, however, you may encounter a maze of unattached wires that add up to more than one white, one black, and one green grounding wire. The clue to how the wires are connected is printed on the face of the ballast, but essentially the fixture circuit must go from the hot line in your house system directly to the ballast, then to each of the lamp holders and back to the house system. The fixture itself can be screwed to the front of an electrical box or hang from the wall or ceiling by brackets of its own.

FLUORESCENT LAMP REPAIR CHART

PROBLEM: Lamp lights slowly or not at all.

CAUSES	REPAIR
No power at source.	Replace fuse or reset circuit breaker.
Tube burned out.	Replace tube.
Tube dirty.	Clean with damp cloth.
Tube and ballast incompatible.	Check label on ballast. Use only tubes with the proper wattage.
Temperature in room too low.	Warm room to 60° F. and try the lamp again.
Tube incorrectly seated in socket.	Check for dust or corrosion on the tube pins and reset the tube in its holders.
Starter faulty.	Replace with a properly rated starter.
Ballast burned out.	Replace.

PROBLEM: Lamp blows circuit interrupter.

CAUSES	REPAIR
Circuit overloaded.	Reduce number of appliances on circuit.
Switch short-circuited.	Check circuit for continuity and repair or replace switch.

PROBLEM: Fixture causes shock.

CAUSES	REPAIR
Case short-circuited.	Test circuit for continuity and repair.
Wiring defective.	Inspect for bare wires and repair.

PROBLEM: Lamp flickers or blinks.

CAUSES	REPAIR
Tube is new.	Flickering should disappear after a few hours' use.
Tube pins not making contact.	Rotate tube in its holders; check pins for corrosion.
Tube burning out.	If the ends of the tube are blackening, replace the tube.
Temperature in room too low.	Warm room to minimum of 60° F. and restart unit.
Starter defective.	Replace starter.
Ballast defective.	Check for loose connections; replace if necessary.
Socket loose.	Tighten socket.

PROBLEM: Tube discolored or only partially lit.

CAUSES	REPAIR
Tube nearly burned out.	If ends of tube have blackened, replace tube.
Poor ballast connection.	If tube is black at one end, tighten connections or replace ballast.
Wiring faulty.	Check all wires and repair.
Starter or ballast defective.	If the tube has light at its ends, but not the center, replace the starter.

PROBLEM: Fixture hums.

CAUSES	REPAIR
Ballast overheated.	Check wiring and repair or replace.
Ballast incorrect for fixture.	Check label for ratings and replace ballast if necessary.
Ballast incorrectly installed.	Tighten all connections; be sure wires are correctly spliced together.
Ballast connections loose.	Tighten all connections.
Ballast old or noisy.	If unit vibrates, tighten connections or replace ballast.

PROBLEM: Tube burns out too soon.

CAUSES	REPAIR
Light turned on and off too frequently.	Replace tube or entire fixture with rapid-start type.
Starter faulty.	Replace starter.
Ballast faulty.	Replace ballast.
Wiring faulty.	Repair wires.

HOUSE SIGNALING SYSTEMS

"House signaling systems" is a fancy term that includes doorbells, buzzers, and chimes. All of these operate on their own low-voltage circuits, which are created by pulling 120-volt household current through a step-down transformer. Every signaling system consists of a push button or buttons, a signaling unit such as a bell, buzzer, or chimes, a transformer, and light-gauge, insulated wiring (usually ＃18 AWG).

You can find all manner of bells, buzzers, and chimes to choose from, but they all operate in about the same way. Electricity flows from the house wiring system into the transformer, where it is reduced to between 5 and 20 volts. When the push button is depressed, it closes contact through the entire system and electricity energizes a small electromagnet on the signaling device that pulls a clapper. An interrupting contact in the signaling mechanism causes the clapper to vibrate and make a noise.

THE TRANSFORMER

All transformers are the same, whether they are the kind that your utility company uses to transmit high-voltage current or the type used in home signaling systems. They all consist of a primary winding coil (just a fat wire wrapped around an iron core) and a secondary winding (skinny, low-voltage wire). The primary winding of the signaling-system transformer is connected to the 120-volt house current, but as electricity is transferred through the low-voltage wiring, it is reduced to voltage so small that you can actually work with the system without fear of shock, even if you have forgotten to turn off the electricity (although it is always better to make a habit of turning off the power when making repairs). A doorbell or buzzer requires a transformer rated between 6 and 10 volts, while chimes typically use transformers rated between 15 and 20 volts.

The primary coil side of most signaling-system transformers has a threaded nipple which fits through a junction-box knockout so that connec-

Fig. 146. The transformer can be mounted on the coverplate of an electrical box and will reduce the incoming house electricity to 6 or 12 volts.

tion with the house wiring is as easy to make as you could ask for. The secondary winding side of the transformer has screw-type terminals for connecting the low-voltage bell wires.

REPLACING A SIGNALING SYSTEM

You might replace all, or only part, of a signaling system, depending on which components have broken down or what you want to replace the system with. For example, if only the push button has stopped working, you need only replace that. If you wanted to change your system from a buzzer to chimes, the buzzer and the transformer must be replaced.

Presuming you want to install an entire new system, the procedure is this:

1. Depending on the size of the push button, drill a hole of the proper diameter for the button in the molding around an outside doorframe.

2. Drill a second hole from the cellar ceiling directly up under the doorframe.

3. Fish the button wires through the two holes. The wires can either be separate or form a cable.

If you already have a signaling system, there is no need to drill holes or fish wires. Tie your new wires to the old ones and pull the old wires up from the cellar and out the push-button hole. When the new wires appear, untie them from the old wires.

4. Connect the wires to the terminals on the back of the push button and screw the button in place over the hole in the doorframe.

5. Tack the wires along the cellar ceiling or joists using insulated staples, and bring them as far as the transformer position.

6. The transformer can be installed in a knockout in the side of your distribution panel and wired directly to the panel. More often, its

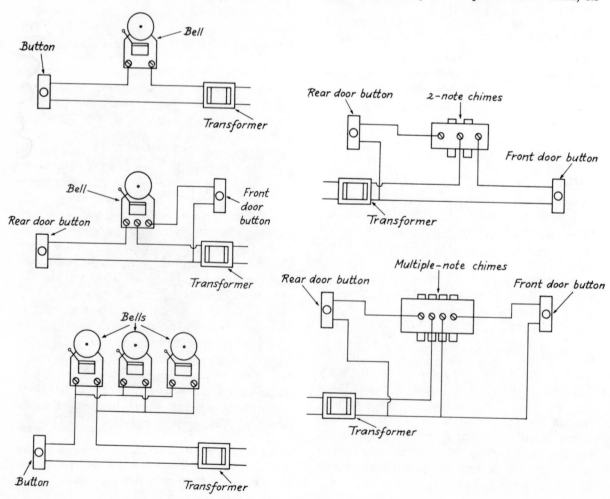

Fig. 147. Wiring arrangements for different house signaling systems.

threaded nipple is inserted through a knockout in the side of a junction box, and is held in place with a locknut.

7. The white transformer wire is spliced to the white wire from the circuit cable entering the box. The black wires are spliced together. The box is then installed to a framing member of the house or on the board that holds the distribution panel.

8. The signaling device must have one wire coming to it from each button installed to control it. Another wire leads from the signaling device back to the transformer. The wires are connected to their appropriate terminals on the signaling device (they are marked) and the transformer. Any signaling system that you purchase comes with a wiring diagram. A few of the more typical wiring arrangements are shown here (Fig. 147).

Fig. 148. What the inside of a signaling system looks like.

REPAIRING SIGNALING SYSTEMS

About 90 percent of the problems that arise in signaling systems stem from the push buttons, because their insides are usually made of iron, which corrodes rather easily. To check whether a bell is functioning, remove it from the doorframe and unhook its wires. Touch the wires together. If the signaling device responds, you have a defective button. You can clean its metal contacts with fine sandpaper, but if the bell still doesn't work, replace it. If, when you touch the wires together, the signaling device does not respond, check for loose wire connections on the device. Also make sure the terminal points are not corroded.

If you know the button is in working order, test the signaling device with a hot-line tester. Place a probe on the tester at each of its terminals and have someone push the system button. If the neon light goes on, power is going through the unit. If the light does not glow, power is not getting to or leaving the unit. Try cleaning all of the wire contacts and any mechanical parts you can reach. If that fails, replace the signaling device.

If the light does not glow, power is not reaching the device, probably because the transformer is faulty. Test a transformer by placing the probes of a hot-line tester against the terminals on its low-voltage side. If the neon light glows, there is a break in the wires between the transformer and the signaling device. If the light does not glow, there is no electrical output from the transformer and you will have to examine the connections in the junction box, as well as the cable from the transformer to the distribution box.

Any time you suspect the wiring in a signaling system to be faulty, look at every splice in the wires very carefully and check the entire length of wire for breaks in its insulation. If you cannot find anything wrong with the wires and the system still does not work, replace the wires.

If a signaling device will not stop ringing, there is either a short circuit in the unit, a faulty push button, or a short in the wire between the button and the signaling device. If the button is not stuck, disconnect it from the wiring. If the ringing stops, the trouble is in the button. If the sound continues, the wiring has short-circuited and must be repaired or replaced.

CHAPTER SEVEN
Upgrading Electrical Capabilities

You need more electrical service in your home if the fuses or circuit breakers in your present system blow or trip with regularity, if the lights flicker or dim when you turn on your appliances, if your appliances do not operate at full power, if the image on your television screen shrinks when you turn on other appliances at the same time, or you are tripping over a viper's nest of extension cords. Actually, if your home has any or all of the above symptoms, you are a member of the majority of American homes, and the only reason you don't encounter endless problems with your house wiring is that you never have occasion to run all of your appliances at the same time.

If you have made a wiring list (see Chapter Two) and determined that your electrical service is inadequate, you do not necessarily have to completely renew all the wiring throughout the building and install a new distribution box. Often, all you need to do is bring in more power from your utility company and add some branch circuits.

SUGGESTIONS FOR UPGRADING

Before you do any upgrading of your electrical system, take a look at your distribution panel to see if there are any unused breaker or fuse positions that would permit the addition of more branch circuits. Even if all the circuit-interrupter positions are filled, remove the face of the panel; you may find a pair of unused terminal screws positioned (usually) between the two left and the two right fuses. These are power-takeoff ter-

minals that can be used to connect an add-on panel box, giving you positions for four new circuits.

If all of the circuit positions and the takeoff terminals are in use, you will need to increase the size of your distribution box after you have added branch circuits to your wiring system.

Call your local utility company and find out whether they are capable of increasing your electrical service. Utilities are delighted to sell you more electricity, but they may not have the equipment, or the capability, of giving you what you want. If, for example, a lot of people in your neighborhood have recently upgraded their homes, the utility may be taxing the transformers in your area to their limit. Or, it may be necessary to increase the size of the cables leading to your home, or any number of technical reasons that can prevent the company from providing the added power your new branch circuits would require. In any case, many utilities are willing to send a representative to your home who is qualified to evaluate your present and future electrical demands and who might even help you plan to meet those needs.

Also learn about your local electrical codes. Some areas allow you to do all your own electrical work. Others permit you to do everything up to, but not including, installation of the panelboard, which must be installed by a licensed electrician. If that is the case, you can pull your own cables and connect them to all the electrical boxes in the system, then pay a licensed electrician to check them over for a fraction of what it would cost to have him do the entire job. That will give you the security of knowing that you have installed a safe, working system which will probably pass official inspection. But even if you do not go to the minor expense of having your work checked before it is examined by the local electrical inspector, get a copy of the electrical code that prevails in your neighborhood and follow its rules and standards.

PLANNING BRANCH CIRCUITS

It is better to have too many circuits than too few (if your original electrical system had been given a generous number of circuits you wouldn't have to add more now). Circuits for lighting and convenience receptacles should generally be kept separate from appliance circuits. By and large, allow each 20-amp, 120-volt circuit to serve no more than 500 square feet of living area, or one 15-amp, 120-volt circuit to cover no more than 375 square feet. Outlets should be spaced between 7 and 12 feet, except in the kitchen, where there should be at least one outlet for every 4 square feet of counter space.

Typically, ⚡12 gauge wire is used throughout a residence, except for major-appliance circuits that feed large-capacity appliances such as ovens and clothes dryers. Number 12 wire allows for 20 amperes or less; if you choose ⚡14 wire, it can only be fused at 15 amperes.

If you are running cable across an open space, such as the ceiling of your basement or the joists in your attic, the NEC requires that it must be supported at least every 4½'. The support comes in the form of special squared staples or plastic

straps. When a cable is exposed in an attic or basement you can protect it in several ways. If it is running along the side or bottom of a joist, rafter, or stud, it need only be held against the surface of the framing member. If it must cross the space between rafters, either lead the cable around the edges of the framing members, or thread it through holes bored at the center of each member.

ADDING ON A DISTRIBUTION PANEL

Small add-on distribution panels can be connected to your main distribution panel and serve as the distribution board for four new branch circuits, controlled by either circuit breakers or fuses.

The two power-takeoff terminals on your existing distribution panel accept the black and red "hot" wires in a 3-wire ⚡12 cable. The white cable wire connects to the neutral bar bonded to the main panel, along with all of the other white circuit wires. The bared grounding wire in the cable connects with all the other grounding wires entering the panel.

The 3-wire cable enters the add-on box and its white wire is connected to the neutral bar. The red wire is attached to two of the fuse positions,

Fig. 150. An add-on panel can contain circuit breakers as well as fuses.

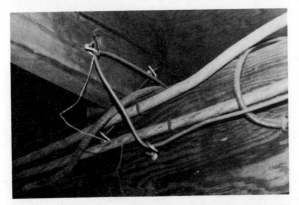

Fig. 149. Electrical cable can be run along joists or through holes drilled in the wood, but it must be supported by some means every 54".

while the black wire feeds the other two. The cable grounding wire is connected to the grounding bar. The branch circuits are then brought into the add-on box. Their hot lines connect to the fuse positions and the white ones attach to the neutral bus bar, while the bare grounding wires connect to the grounding bus bar.

WIRING FOR SPECIAL EQUIPMENT

There is available to every homeowner an almost endless array of special electrical equipment, some of which can affect the conservation of power, much of it necessary for everyday comforts. Many of the appliances must have their own special branch circuit. Many others need only an outlet they can be plugged in to.

INSTALLING EXHAUST VENTILATORS

Powered kitchen ventilators are required by some building codes to carry smoke and odors from a kitchen to the outside. The ventilator is normally installed in the wall behind (or the ceiling above) the cooktop or stove, and is positioned at the end of a metal duct that leads through the wall or ceiling to the outer skin of the house. The vent itself is a fan consisting of four deep-angled blades attached to the shaft of a shaded-pole motor that typically operates at one speed only, although variable speed models are available.

Exhaust fans, whether they are used to pull air from the living spaces up to the attic, or are situated inside a housing perched on the slope of your roof, or are buried in the wall or ceiling of your kitchen, do not need to be on a special branch circuit. They are sold with a sleeve of some type that is secured to the sides of whatever hole is made in the wall or ceiling to hold the fan.

Some exhaust fans have a line cord and plug, in which case you plug them into the nearest outlet. Others have lead wires from their motors which are connected to black and white branch cable wires entering a junction box. You can, of course, splice the motor leads to a cable (with wire nuts) and then run the cable to the nearest light fixture or even outlet or switch box and tap into the source of power at that point.

Fig. 151. The ventilators over a stove contain a fan to draw warm air and odors out of the building, and may also have a work light.

Fig. 152. A powered roof fan can be wired into an existing light circuit.

Considering that the exhaust fan may need to be at the end of a duct, you will have to do more carpentry work than electrical work when installing one.

INSTALLING A GARBAGE DISPOSAL

Garbage disposals are canisters attached to the drain of a kitchen sink and then linked to the sink trap and ultimately the house drain-waste-vent line. The plumbing is not very complicated since it can be done with metal or PVC plastic pipe kits that are readily available at most hardware and plumbing supply stores. Moreover, the actual installation of a garbage disposal is detailed in the installation sheet that comes with the unit.

Disposals are motor-operated devices. The motor is at the bottom of the canister and care-

Fig. 153. A garbage disposal is hung under the drain of a kitchen sink.

Fig. 154. The disposal motor is situated at the bottom of the canister and must be wired to a nearby switch.

fully isolated from the cabinet at the top of the unit that receives both water and garbage. Only the motor shaft protrudes up into the grinding portion of the unit, and the leads from the motor are reached by removing an access panel on the bottom of the appliance.

The electrical hookup for a garbage disposal must be at least a 15-amp circuit controlled by a standard wall switch. Since the disposal is normally operated for only a few minutes at a time, the most practical approach to providing power is to tap into an existing 120-volt, 15-amp or 20-amp branch circuit. You can do this by splicing the black and white motor leads to a cable (the splice remains inside the motor cavity) and running the cable from under your sink to the nearest wall containing a receptacle. However, before you get to the receptacle or junction box, install a standard single-pole toggle switch in a switch box. The switch should be within 50′ of the disposal, and well in sight of anyone using the appliance. One place to install it is on the wall of the cabinet under the sink. The cable coming from the disposal can be cut so that the two black wires can connect to the switch. Splice the white wires together inside the switch box and continue on to the box where you are connecting the cable to the branch circuit. People who use garbage disposals are likely to have wet hands when they touch the toggle switch; not only should the circuit be completely grounded, but it should also have a GFCI controlling the circuit.

As with fans, the installation of a garbage disposal causes you to function more as a different kind of handyman than an electrician. In this case, you will do more plumbing than wiring.

INSTALLING BATHROOM HEATERS

It is an unusual fact of house heating systems that the bathrooms often wind up at the end of the heating line, farthest from the furnace. Bathrooms are often the last rooms to warm up when the furnace goes on. The simplest way around this particular aggravation is to install an electric heater in the bathroom wall or ceiling.

Ceiling-mounted heaters can range anywhere from 600 to more than 1500 watts of power. Some models include a heater and overhead lamp, others incorporate a fan, and some have a timer switch that can vary the length of time the heater is on from 60 seconds to an hour. Whatever time period you set the switch for is the length of time the heater will be on, and then it will automatically shut off, which is a fine energy-saver. People often remember to turn off the bathroom light when they leave the room but not the heater, so a timer switch can control the amount of money you spend heating the room electrically.

The ceiling fans used in bathrooms often (but not necessarily) incorporate a light. The heaters can deliver enough heat so that the unit must be connected to its own 240-volt line, although for most bathrooms the unit used can be operated from a 120-volt, 15-amp or 20-amp circuit.

Installation of the heater is a carpentry job involving a hole cut in the ceiling or wall between joists or studs (Fig. 155). The hole is framed with lumber and the heater is installed once the wiring has been completed. The wiring to the heater is normally a cable brought directly to a junction box attached to the heater housing. The black heater wire connects to the black cable wire, and the white heater wire connects to the white cable wire; in the case of a heater/light, you have separate wires from the heater and the light. Manufacturers usually provide wiring diagrams with their products which, if you follow them, not only make the unit work properly, but also meet NEC regulations.

The timer switch fits into a standard switch box and is wired to the heater, or to both the heater and the light, in the same manner as any switch. That is, one line in the 2-wire cable is spliced to a hot line from the heater and the white (painted or taped) wire is spliced to the black cable wire. If the light is controlled by a separate switch, this is normally a separate toggle switch positioned on the wall next to the timer and is wired separately to the light. If there is no wiring diagram with the unit, or if you want to verify the wiring arrangements, see the wiring

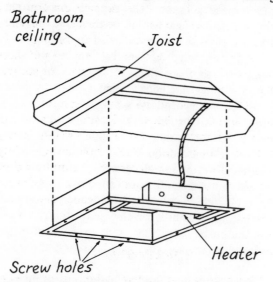

Fig. 155. The housing for ceiling fixtures and overhead heaters is fitted into a framed hole and held in place with screws.

diagrams in Chapter Four. You can replace any of the switches in the diagrams with timer switches; the wiring arrangement will be the same.

INSTALLING HOME CONTROL SYSTEMS

Home control systems are a conglomeration of electronic components that provide remote control of the lights and different appliances throughout your home (Fig. 156). The best part about them is you don't have to install any special wiring.

The entire control system is run by a command console that is plugged into any 120-volt AC branch circuit and will control lamps, ceiling fixtures, and outside lighting, as well as TV sets, hi-fi systems, radios, fans, dehumidifiers, coffeemakers, garage-door openers, heating cables, and any other appliance you might want to turn on, off, or dim from a central position.

The command console is used in conjunction with wall switch, appliance, and lamp modules which must be plugged in to a house outlet. Some systems also provide a cordless controller. All of these modules are manufactured using miniaturized electronic computers (known as microprocessors) and solid-state circuitry, so they can be relied on for dependable service. In fact, if anything goes wrong with the unit you cannot repair it but must return it to the manufacturer for replacement of the circuitry.

Command console. This is a small box containing a variety of buttons. The buttons are unit keys that correspond to the unit code dial on the back of each lamp, appliance, and wall switch module, so that by pushing a given button, a particular lamp or appliance will be activated no matter where you happen to be in your home. All of the signals from the console are transmitted through the house wiring.

Lamp module. The lamp modules have plug prongs in their backs so they can be plugged in to any outlet. Lamp modules are designed to control incandescent lamps, not fluorescents, and can handle up to 300 watts. You can plug one 300-watt lamp in to the module, or several lamps, so long as their total wattage does not add up to more than 300. Once the module's dial code has been set to an identical code number on the con-

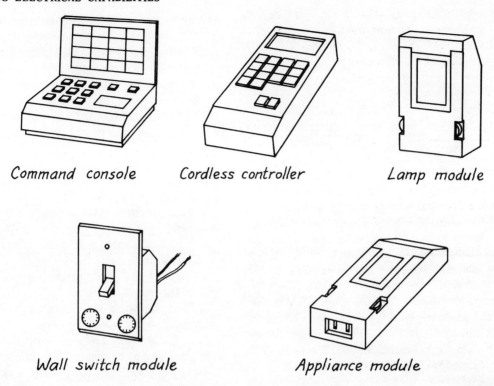

Fig. 156. Home control system components.

sole, it can be turned on or off from the console, as well as dimmed or brightened.

Appliance module. These are also plugged in to any outlet in the house and their dials are set to the corresponding numbers on the control console. Then the appliance is plugged in to the module so that the console can control it. Fluorescent lamps, incandescent lamps, fans, humidifiers, television sets, coffeemakers, practically anything can be plugged into an appliance module so long as its wattage does not exceed the rating of the module.

Wall module. The only wiring necessary with home control systems occurs when you elect to install ceiling fixtures with a wall module. The module has a standard toggle switch and fits into any switch box with its lead wires spliced to the black wires of the switch cable entering the box. Once the module is installed, it can be dialed to one of the channels in the console and from then on be used either directly or via the control console. The wall modules have ratings of up to 500 watts, so they are particularly useful for outdoor wiring systems.

Cordless controller. The cordless controllers are battery-operated ultrasonic transmitters. They have a keyboard identical to the command console. The drawbacks to the unit are that they must be pointed directly at whatever you want to turn on or off, you have to be no farther away than 30′, and they can't activate anything if there is a wall between you and the device you are trying to control.

The home control systems are sort of fun to have around the house and they certainly will give you remote control of all the lighting in and around the building. Such monitoring ability should give you the advantage of turning off or dimming lights and shutting down appliances more readily, and therefore conserving electricity.

CHANGING THERMOSTATS

The heating and cooling systems that provide comfortable temperatures in your home are all controlled by either thermostats and/or humidistats. Thermostats are available in several types

Fig. 157. Some thermostats include a clock and can be programmed to activate the furnace at different times of the day and night.

and designs, some of which are fairly sophisticated and complex. Most, however, operate on the principle of a bimetal strip or disk which expands or contracts to open or close contacts that complete a circuit to the furnace or central air conditioner. Most home thermostats operate on about 24 volts, which means the wires connecting them to your furnace or A/C are #16 or #18 low-voltage. Like home signaling systems, the thermostat will not give you much of a noticeable shock, but it still is a smart idea to turn off the current in the branch circuit serving the furnace or air conditioner before you tamper with the thermostat.

Fig. 158. The wiring that feeds a thermostat is usually low-voltage bell wire.

Thermostats wear out in time and must be replaced, and when you get around to making that replacement, you have a full range of units to select from. You can purchase a basic, manually set unit, one with a clock in it, automatic versions that can be preset to different temperatures for day and night cycles, and still others that allow you to program the temperature for different times of the day and night through a

full seven-day cycle. The contacts and assembly of the wiring may vary slightly with the type and model thermostat you select, but the general procedure for replacing a thermostat is this:

1. Remove the fuse or shut off the circuit breaker controlling the thermostat.

2. Remove the faceplate of the old unit and remove the mounting screws holding it to the wall.

3. Pull the old thermostat away from the wall. Loosen the terminal screws on its back. It is very easy for the loose wires to fall down into the wall; wrap them around a pencil so you don't lose them.

4. Scrape the bared wire ends with the back of a knife until the wire is shiny.

5. Connect the wires to the terminals on the back of the new thermostat.

6. Push the wires back into the wall and put one mounting screw into the wall.

7. If the new unit has a mercury vial in it, use it as your guide for making the unit level. Thermostats *must* be level, so if there is no mercury tube, place a small level on top of the thermostat and level it.

8. When the thermostat is absolutely level, install the second mounting screw.

9. Put the face of the thermostat on the unit, and turn on the electricity.

10. Turn the thermostat up and down to make certain it is activating the heating or cooling system.

INSTALLING A ROOM AIR CONDITIONER

Some modern room air conditioners are made to operate on 120-volt circuits. If that is the type you own, you can plug it in to practically any circuit in your house. Most large air conditioners require their own branch circuit providing 240 volts of power. When the unit is in operation, the voltage must be within 10 percent of the unit's rated voltage, and you must have the proper circuit interrupter with the correct amperage rating controlling the line. Exactly what wire size and which ampere rating is used for the fuse or circuit breaker depends on the machine you are installing, but it will be something like a 3-wire #12 cable bringing in 240 volts at 30

Fig. 159. House air conditioners are larger and more powerful than the room units. They must be placed outside the house and given their own separate branch circuit.

Fig. 160. House intercoms are really one-way telephones containing an amplifier, loudspeaker/microphone, and controls.

amperes. Even more importantly, the outlet installed at the end of your branch line must have the proper design of receptacle to accept the grounding contact on the unit's plug. The branch line serving an air conditioner should have no other appliances on it, and above all must be completely grounded.

INTERCOMS

Intercom systems in a home can function as step-savers, voice-savers, security systems, and even as babysitters. The simplest type of intercom system includes a master unit which contains a voice amplifier, a loudspeaker which also operates as a microphone, and some control knobs. There is also a remote box which is just a loudspeaker. The system may be battery-powered, in which case it is eminently usable anywhere in or around your house. The components may also have to be plugged in to a standard 120-volt branch line receptacle.

Unlike a telephone, you cannot talk back at will on an intercom system. The control unit must be used to initiate a call by pressing a button. Releasing the "talk" switch allows whoever is at the remote box to speak. You can, however, tie as many remotes as you wish to a single control.

The one-way communication provided by in-

tercoms is ideal between your front door and, say, the kitchen. But you can also leave your intercom system on all day long (they demand about as much electricity as an electric clock) and the remotes, even in the cheapest models, can pick up voices that are as far as 50′ away from the unit. So you can be upstairs in your bedroom and still hear your kids playing in the basement, or even the backyard.

While intercoms draw their power from small 9-volt batteries or house current, they nevertheless have to be wired to each other. The wires are small and easily concealed, but they still have to be either pulled through the walls or artfully stapled up the side of door molding and in and out of nooks and crevices, and that can be a tedious chore at best.

There are also "wireless" intercoms which are merely plugged in to the house wiring and use branch circuits to carry your voice all over the house. Essentially, each master and receiver unit is a small radio transceiver, i.e., they each contain both a transmitter and a receiver. All of the units in a particular house system are tuned to a frequency that is below even AM broadcast bands, so they will not interfere with any equipment in the house. The wireless equipment, since it contains more complicated components, tends to be more expensive than standard intercom systems. But then, they are more versatile because they can be moved around and plugged in to any receptacle in the house and instantly be put in service.

TRACK LIGHTING

Track lighting is a series of light fixtures suspended from an electrified track attached to the ceiling. The fixtures can be moved back and forth along the track and are usually of the spotlight variety, which means they can be swiveled in all directions to create an almost endless variety of lighting effects.

Traditional track systems are expensive, since you are buying not only a number of light fixtures, but also the track that holds them and the wiring inside the track to power the lights. But there are now available several plastic versions of the standard metal-track systems. These are less expensive and come with a single power cord that is plugged in to any outlet in the house. Obviously, these are easier to install than the normal tracks.

The metal tracks are sold in 4' lengths that are connected together and installed on your ceiling. Since the tracks must support a considerable weight, either you want to follow under a joist and drive your mounting screws into that, or, if you are crossing joists, you want to screw into each one of them. But the tracks are the least of your worries; every light in the system must be controlled by a wall switch somewhere. Moreover, if you have a lot of lights, and/or they are high-wattage units, you may need to tap into more than one branch circuit, or run a separate circuit from your panel box to the track.

Track systems, fortunately, come with wiring diagrams, but you can still get into a tangle of wires coming out of the track, disappearing into the ceiling and walls, and finally emerging at a bank of switches. To make the installation even more fun (and the lighting more versatile), many if not all of those switches ought to be dimmers. The switches (or dimmers) are all wired in the standard way (see Chapter Four); it's just that you will have a lot of them to install. Don't plan on finishing the wiring in a single day, as it could take you a couple of weekends to finish the job. Above all, be careful not to tap into already loaded branch circuits or you will be blowing fuses every time you light up the track.

INSTALLING A GARAGE-DOOR OPENER

An electric garage-door opener is one of those pieces of electronic equipment that you can live without and never notice the difference it makes in your life—providing you have always had to live without one. Install one on your garage door for the first time and you will never be without one again.

You are back to being a mechanic, rather than an electrician, when you decide to install your own garage-door opener. But even before you buy the unit, be sure about several things. First, make certain of the type of door you have. Most electric openers will work with doors that slide on either straight or curved tracks, but you want to be sure you are buying a unit best suited to the kind of door on your garage. You also have to be aware of the *size* of the door, since its weight will affect how powerful an opener you purchase. If you have a cast-iron, two-ton door, don't expect to open and close it with a half-horsepower motor. Also check the distance from the highest arc of the door and the garage ceiling. The opener must have enough space above the door to have about 2" of clearance.

Garage-door openers consist of a track that contains a chain-driven connecting arm which is attached to the top of the garage door. One end of the track is bolted to the header above the door and the other end of it holds the motor, clutch wheel, switches, and wires. The motor is wired directly to a 120-volt, 15-amp circuit, or it may have a plug attached to it. It is also wired to a wall switch inside the garage and a control

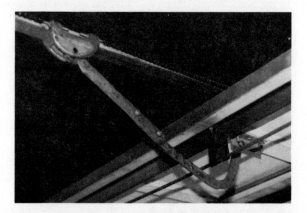

Fig. 161. The garage door is attached to an arm that travels along a track. The arm is pulled by a chain.

Fig. 162. The opener itself consists of a motor that is activated by radio signals but is wired directly to a house circuit.

box, which these days is a solid-state radio receiver. The transmitter for the control unit is a small battery-powered unit which you carry in your car. The transmitter, or "genie" as they are sometimes called, has a button on it which, when you depress it, sends a signal to the control to activate the opener motor. The motor rotates a chain attached to a connecting arm and pulls the arm and the door along the track.

Wiring the garage-door opener is an anticlimax to hanging the track and motor. With the exception of the cable from the branch circuit, all of the wires are small (⚡18 gauge or so) and are attached to plainly marked terminals on the back of the motor housing. A 24-volt wire/cable runs from the wall switch to the motor housing, and another 2-wire cable comes from the radio control unit to the same terminals. The radio control

Fig. 163. The wiring hookup amounts to connecting the button wires to terminals and connecting the branch circuit cable to the motor.

unit is normally plugged in to a light socket fitted with a receptacle or to any handy receptacle. The cable from the house branch circuit is clamped to the motor housing as it enters the housing; its wires are attached to screw terminals inside the motor housing.

ALARM SYSTEMS

Every security alarm system consists of sensors, a control unit, and the alarm itself. These can be combined into a single unit, although you have considerably more flexibility when designing a system if you use separate components.

SENSORS

The function of the sensors is to detect any uninvited guests coming into your home and send a signal to the control unit, which in turn triggers the alarm. There are several types of sensors, each of which operate in different ways.

Switch type sensors are electromagnetic devices installed at doors and windows so that when the door or window is opened the switch automatically closes a circuit to the control box.

Pressure mats are really large switches hidden under the carpet at entranceways, or in a hall, or near a valuable item. As soon as anyone steps on a pressure mat the control box receives a signal to activate the alarm.

Ultrasonic motion detectors are placed in a room so that when they are turned on they fill the area with sound waves that are too high to be heard by the human ear (something over 40 Hz). Any movement in the room, however, will disturb the sound wave pattern and trigger the alarm.

Infrared photoelectric sensors project an infrared beam between two points, such as across a hall doorway. As long as no one breaks the beam, all is quiet. But as soon as it is interrupted the alarm will go off.

ALARMS

There are two types of alarms, local and remote. The remote alarms send a signal to some place away from the home, to a neighbor's house, for example, or your office, or the local police station, or a private security agency. Local alarms are easier to install and are normally hidden in the attic or under the eaves of the house. They are loud bells or sirens that are meant to scare an intruder away from his self-appointed rounds on your private property and/or alert your neighbors to call the police. Years of research have shown that a loud, on-the-spot alarm will deter more attempted entries since the noise increases the possibilities of discovery and arrest during a crime.

WIRING

Electronic security systems can be what is known as hard-wired, line-carrier, or wireless types. The hard-wired systems use their own wiring circuit and batteries to transmit signals from the sensors to the control box and alarm. Line-carrier systems rely on house wiring to transmit their signals. Wireless systems use infrared, microwaves, radio frequencies, or ultrasonic sound waves.

When you start shopping for an alarm system you will encounter a multitude of choices, not only among the basic components, but in the scope and versatility of the system you can install. You can assemble your own system with components made by different manufacturers, or purchase complete, packaged systems. You can also hire a commercial company to install your system for you, or you can do it yourself. It goes without saying that if you take on the job yourself, whatever components you use will come with complete installation instructions and wiring information, all of which should be followed exactly.

When planning your system you want to provide adequate protection at all ground-level doors and windows and any other place large enough for human entry into your home. "Ground level" means any access area within 12′ of the ground, but don't forget balconies, upper-story windows that are near tree limbs, and the roof if you have a roof hatch. Each of these areas should have a switch sensor protecting it that is wired to the control box and alarm.

When you have planned for sensors completely around the perimeter or exterior of your home, don't stop. A sensible system also includes bug-

ging any logical area inside the building where an intruder will have to travel, a central hallway for example, the top or bottom of a stairway, and near valuable possessions such as your television, stereo set, or silver closet.

INSTALLATION SUGGESTIONS

Because every home is different, every alarm system will also be different, and the various components used will be installed a little differently as well. There are some general suggestions that might be kept in mind, however:

1. The tools you will need to install an alarm system include a wire fish, power drill, screwdrivers, U-nails, silicone cement, wallboard joint compound, plastic wood, and a soldering iron. The wire connections at the sensors and control box are often soldered.

2. The master alarm should be placed where it cannot be seen or reached. This often means the attic or under the eaves of the house, if the eaves are more than one story above the ground.

3. All wiring from the sensors to the control box or the alarm must be carefully hidden along moldings and baseboards, or pulled through the walls. Hiding all of the wires is not a matter of aesthetics as much as it is to make the alarm system invulnerable to being deactivated, since a visible wire can easily be cut. The best approach is to run most of the wiring through your basement or attic and keep the wire run through your walls as short as possible.

4. The control box can be placed in your basement stairwell. If you are installing a line-carrier system, it should have an auxiliary power supply and the ability to automatically convert from AC line current to DC battery power if there is a power failure in your house.

5. If you mount an alarm on the outside of your house, be certain that none of the wiring is exposed. Outside alarms should also be weatherproof and have a tamper-proof siren or bell box that will activate the alarm as soon as anyone tries to disconnect it.

6. Remember that double-hung windows and dutch doors have two moving parts. Each half of the window or door must have a sensor protecting it.

7. You can connect what is called a panic but-

ton to almost any alarm system and put it in your bedroom or other convenient place in the house. The panic button is just that: when you wake up in a cold sweat in the middle of the night because you think you heard someone sneaking around downstairs, you can hit the panic button and your alarm will immediately wake up everybody else in the neighborhood.

CHAPTER EIGHT

Maintaining Home Appliances

All appliances that require some form of electricity to operate contain their own electrical circuits. The circuits are comprised of many different components, some of which have been engineered for use only in the unit where you find them, and many of which are common to all appliances. Most electrical equipment has an on-off switch; many require a motor of some kind in order to operate; timers, heating elements, thermostats, and relays are common to many others. Consequently, even though a component such as a timer or thermostat may look different from all the other timers and thermostats you ever encountered, they all function in basically the same manner, and usually they are repaired with the same basic procedure. But to be honest, the great majority of all appliance repairs do not involve repairing a part. They involve replacing one.

MANUALS, WARRANTIES, AND PARTS LISTS

When you buy even the smallest, least-complicated electrical appliance, you almost always receive a warranty card, an owner's manual, and often a parts list. The warranty card must be signed and mailed to the manufacturer if it is to be put into effect; typically, it promises the purchaser that the manufacturer will replace defective parts or correct poor workmanship without cost to the owner during a specified period of time, such as one year. During the period when the warranty is in effect, should anything go wrong with the appliance, by all means return it to the manufacturer for repair or replacement. After all, it's free.

Owner's manuals tell you how to operate your appliance and, when appropriate, how to maintain it by keeping it clean and perhaps lubricating some of its parts once or twice a year. It is a good idea to keep all your owner's manuals for as long as you have the appliances, not only because of the various instructions in them, but also because they may include a parts list. The parts list is often accompanied by an exploded drawing of the appliance that shows each of its parts. The list then states the name of each part and a number, both of which you will need should you ever have occasion to write the manufacturer in quest of a component.

Service manuals do not normally come with appliances but must be requested of the dealer or the manufacturer, and often have a price tag attached to them. A service manual for any appliance, particularly for the large appliances such as dishwashers or refrigerators, is absolutely indispensable if you intend to make any repairs on the machine. The service manual explains all the mechanical parts, where they are located, and how they can be repaired or replaced. It also describes the electrical circuits. Even a professional repairman must have the manual for whatever machine he is working on, because while many of the components may be identical, they may be wired differently or installed in an unusual manner.

As a matter of course, homeowners purchase a service manual whenever they buy a new appliance, and just about every manufacturer has some sort of service manual for every model of every piece of equipment he makes and sells. Given the service manual for an appliance, plus the ability to read and follow instructions, you would think just about anybody could repair almost any electrical equipment on the marketplace today. And so they can, providing the manual has been clearly written, which many of them are not. On the other hand, service manuals are filled with diagrams and pictures, and they do contain all the information necessary to effect any repair on the machine they describe.

PURCHASING PARTS

The major appliances, washing machines, clothes dryers, dishwashers, stoves, ovens, and air conditioners, are marketed by scores of manufacturers, and each manufacturer is likely to produce several models of each appliance in his line. Each appliance may include hundreds of parts, any of which could require repair or replacement at any time. But the major appliances still number only a relatively few types and account for comparatively few parts. Consequently, there is probably a major appliance parts store within fifty miles of your home, wherever you happen to live.

Appliance parts stores carry many of the components needed for any major appliance on the market; even at that, however, they cannot stock every part for every appliance. But they can order them from the manufacturers, often getting delivery within a day or two. So, if you have a major appliance in need of new parts, head for the nearest appliance parts store. If there is no such retail outlet listed in your yellow pages, your alternative is to communicate directly with the manufacturer and request the part you need. This is where the parts list in your owner's manual becomes invaluable. The best (sometimes the only) way of ordering replacement parts is by stipulating the model name and number of the appliance, together with the part name and number.

Small appliances, which include everything from radios to all of the portable gadgets in your kitchen, represent thousands of different types, models, and styles of machinery, and literally millions upon millions of components. Most of the parts are small and inexpensive, so the pure force of their numbers, together with the fact that an average sale would probably be counted in nickels, has prevented retail dealers from going into the replacement part business. True, many large manufacturers maintain rehabilitation centers for their line of products, but many do not. Most manufacturers do have a local appliance repairman representing them in major areas around the country. But the cost of parts and labor, and the demands of our economy in general, can run the cost of, say, replacing the warming plate on your old coffeemaker to just about what it would cost to buy a whole new appliance.

If a small appliance goes on the blink and you can figure out what is wrong with it, the chances are better than even that you can also fix it. Particularly with small appliances, most repairs mean replacing a part or two. Replacing a part means you disengage one component from the appliance and install an identical one in its place. You hardly need to know what you are doing to do it right. The rub comes in getting the right replacement part.

If the manufacturer of your ailing small appliance has a repair outlet near you, discern the name and number of the part you need from the parts list (which is in the owner's manual you have been saving for the past five years) and call the outlet. If they have the part you need, then go buy it from them. If they do not have the part, they can get it for you, but it may take a week or two.

If there is no repair outlet near you, you might try some of the repair people listed in your yellow pages as representatives of the manufacturer. Once in a great while the repair shops do sell some parts, especially the ones that have proved to be in demand. The warming plates in drip coffeemakers, for example, are vulnerable to burning out and often you can buy replacements at local repair shops.

When all else fails, write to the manufacturer. State the name and number of the part you need, and include the broken part if you can. The manufacturers are, by and large, very gracious about sending you components, and sometimes even instructions for installing them. They are gracious to the point that often you are not even charged for your purchase. But, consider that you wait a week for your request to reach the manufacturer through the mails, another week before the part is shipped back to you, and a third week for it to be delivered to your home. So the appliance is out of commission for nearly a month, presuming the mails do not falter and the manufacturer's shipping department responds to your request with a modicum of alacrity.

HOW TO MAKE APPLIANCE REPAIRS

When an appliance stops operating, don't blindly begin taking it apart. There is actually a step-by-step procedure to follow that may well help you avoid aggravation, headaches, and hard labor.

1. Double-check the power cord and plug. The plug should be properly seated in its wall socket. But also look at the cord for any damaged insulation or exposed wires.

2. Check the house branch circuit. The fuse may have blown or the circuit breaker snapped to its "reset" position.

3. If the appliance uses water, be sure the faucets on the supply line feeding the unit are open and that water is reaching the unit.

4. Once you are sure the appliance is receiving the necessary electricity (and water) and the line cord is in order, unplug it and begin to think about taking it apart.

5. Look at all the dials, knobs, and buttons. Work them to be sure that each one is operating mechanically. Clean them of any grease or dirt.

6. Turn the unit over; examine its shell closely for any sign of cracks, breaks, chips, holes, anything that might be affecting its normal operation.

NOTES ON DISASSEMBLY

If you have discovered nothing untoward up to this point, you can begin your disassembly of the unit. But do it slowly and deliberately. Every part you remove is a part you will have to put back in place, and if you have too many of them lying around on your workbench, you may forget exactly how each one must be situated in the appliance.

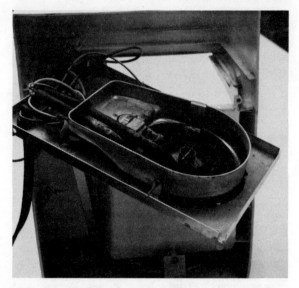

Fig. 164. The inside of your coffeemaker is nowhere near as complicated as a toaster, but you still have to make sure every wire is connected to the right place after you have made your repairs.

Open the shell of the appliance, and before you touch anything inside, either make a sketch of what you see or, better still, take a Polaroid picture of it.

When you remove a component, note exactly how it aligns with the parts around it. You can label parts with bits of masking tape, or mark their alignment with a felt-tip pen or sometimes even a pencil. Keep all nuts, bolts, springs, washers, screws, any tiny parts, in separate, labeled containers.

Never use excessive force to get parts apart (or together). All of the parts in an appliance have been engineered to very close tolerances and they are supposed to fit perfectly with the components around them. If a part appears to be "frozen," look for a hidden fastener. If the parts are metal, coat them with a penetrating oil and wait a few minutes, then try to work them free. Some metal and plastic parts are force-fitted together, but they will come apart if you very carefully apply heat to them. Use a soldering iron and touch its tip to the *metal* part until the metal is warm enough to expand, and also to soften the plastic slightly. Then gently pull at the parts. Go slowly with this procedure, and apply your heat only a little at a time, working the pieces while they are still warm.

When you have completed your repairs, check all of the parts in the appliance. Clean all electrical contacts and be sure that each of the switches is working properly. As a matter of course, check the shaft on any motor for looseness or wear, and oil the motor if it is the type that requires lubrication. Check all the wires, looking for any frayed insulation and loose or dirty connections.

Put the parts back in the appliance housing in the reverse order you took them out. Do not crank any nuts or screws down so tightly that you strip their threads. Any part that you can move by hand should be manually tested to be sure it is not binding against anything.

When you have gotten all of the components back in place and have reassembled the shell around them, turn on the appliance. Be sensitive to any unusual noises, smells, or heat.

BASIC PARTS

As mentioned before, certain parts are common to many or all appliances, and while they may look somewhat different from each other, they function in the same manner. They can be installed in different appliances in a multitude of ways, so it pays to look carefully at exactly how a part has been installed, what other parts it touches and how, what angle it is positioned in, and so on. Your job as a repairer is to get the replacement for any part in exactly the same position, and with precisely the same wire connections, as the component you are replacing.

PLUGS AND CORDS

Every appliance (unless it runs on batteries) has a plug or a cord, and most have both. Power cords and plugs are a common source of problems and should be about the first component that you look to for trouble.

Fig. 165. Female plugs have receptacles for the prongs on male plugs.

There are two kinds of plugs, male and female. Male plugs have prongs that are meant to fit into the slots in wall outlets and/or female plugs. The female plugs are often found on one end of a removable cord such as is used with some coffeemakers, electric skillets, and other kitchen appliances. The female plug may be a set of parallel slots or holes in the end of a molded plastic head. It may also be a shell held together by screws. If the plug can be dismantled, the cord attached to it can be removed and replaced without cutting the wire.

Power cords are 2-wire cables and you can buy the wire for them at any electrical or hard-

Fig. 166. The wires feeding both male and female plugs are installed on terminals that are usually screws.

Fig. 167. The strain relief device may take any of several forms but essentially it attaches to the shell of an appliance and prevents the cord from being pulled loose at its connections.

ware store. Be certain that the cord you purchase contains wires of the same gauge as the original, and that it has a grounding wire if the original had one. When you are making up a replacement cordset, either use the strain-relief device on the appliance to protect the wires from being loosened from their contacts, or tie an Underwriters knot both in the plug and on the inside of the appliance end of the cord.

When replacing an appliance cordset, it may be necessary to dismantle the appliance enough so that you can open the splices or terminals that connect the cord with the internal electrical system, but this is usually no problem at all.

LEVELING LEGS

Leveling legs are commonly found on washing machines, dishwashers, refrigerators, clothes washers, dryers, and stoves. The legs are normally threaded pipes with large heads that can be screwed up and down to different heights to compensate for an uneven floor. Many large appliances will not operate efficiently unless they are sitting level in all directions. A washing machine, for example, will wobble horrendously (and make a great deal of noise as well) unless it is absolutely level.

To check the level of an appliance, place a carpenter's level on top of the cabinet so that it is going from side to side. Then read the level while it is going from front to back. Adjust the appropriate leveling leg(s) so that the bubble in the level rests between its centering lines no matter which way you place the level.

When leveling an oven, place your level on the cooking racks, not the top of the unit. Refrigerators and air conditioners should be level from side to side, but tilt slightly backward so that condensation can drain out of the appliance. The front (inside) of an air conditioner should be 1/4" higher than its back (outside). Refrigerators can also tilt back 1/4".

SECRET FASTENERS

Over the years, manufacturers have devised a whole catalog of ways to hide the screws, bolts, and clips that hold the outer coverings of their appliances together. If you look at an appliance

and see no sign of any fasteners, carefully pry up the corner of any decorative decals such as might be around a row of knobs or dials. There will often be screws or bolts under the corners of the decal (Fig. 168).

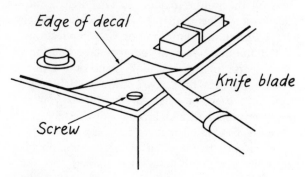

Fig. 168. Sometimes screws are hidden under a decorative decal.

Screws and bolts are sometimes inserted through the feet of small appliances. If you do not see any screw or bolt heads, but there are several circles in the plastic sides of a unit, pry one up with the point of a knife (Fig. 169). The disk is most likely a cap over a hidden fastener.

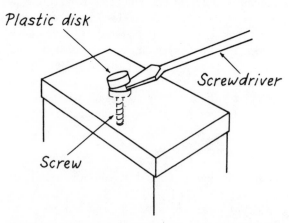

Fig. 169. If there are a lot of disks in a plastic housing, pry one of them up; it probably hides a screw.

Some plastic appliance shells have been assembled using a notch and tab arrangement. The two halves of the shell can be pried apart with a screwdriver or knife, but use caution and do not force the parts.

The metal sides of such large cabinet appliances as dishwashers, clothes washers and

Fig. 170. Some units merely snap together and are held by interlocking tabs.

dryers are often held in place by metal clips hidden between the tops of the side plates and the top of the cabinet. The spring clips are normally at the corners and are released by sliding the blade of a screwdriver under the top and pushing it against the spring while pulling the side away from the frame of the unit (Fig. 171).

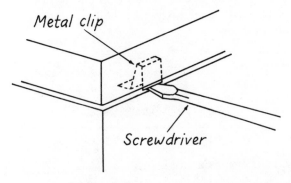

Fig. 171. Large appliances are held together by hidden spring clips.

KNOBS

Most often, before you can disassemble an appliance with any sort of knob or dial on its face, you have to remove the knob. Some knobs just pull off their shafts. Others are held to their shafts with a small setscrew that must first be loosened. Still others rely on a spring clip inside the knob to hold it on its shaft. If you find the end of the clip behind the base of the knob,

depress it with a screwdriver blade and simultaneously pull the knob outward.

Some knobs have a setscrew in their hubs. The screw may be a standard slot screw, a phillips-head, or a hexagonal hole requiring the proper-sized *allen wrench* to loosen it (Fig. 172). There might also be what is known as a driftpin through the hub of the knob. Driftpins are small steel dowels that fit through the knob and a corresponding hole in the knob shaft. To remove one, tap it lightly with the point of a small nail or awl. When you can grip the pin with your pliers, pull it out of its hole.

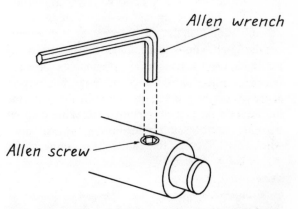

Fig. 172. Allen wrenches are the only way to loosen an allen screw.

If you examine a knob and find no apparent way of holding it to its shaft, work a knife or screwdriver blade under the cap covering the center of the knob. Often, the cap can be pried up to reveal a retaining fastener (Fig. 173).

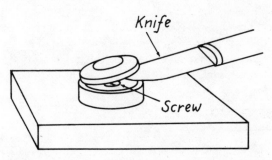

Fig. 173. The caps set in the center of knobs are spring-held and can easily be pried up to get at the knob's setscrew.

WIRING HARNESS

All of the electrically powered components in every appliance are connected by wires. All of the wires (and there may be a score or more of them) are collectively called the *wiring harness*. The harness is actually a series of separate circuits for each function or mode that the appliance can perform. For example, in a dishwasher there are separate circuits for the wash, rinse, hold, and dry cycles.

If a majority of the wires in an appliance are worn and need to be replaced, the simplest approach is to ask the manufacturer for a new wiring harness for the model you are repairing. You will receive a bunch of color-coded wires, all different lengths, held together by plastic bands. Each wire is exactly the length it needs to be to connect the specific components at each end of it and may or may not have connectors crimped to their ends. Install each wire in the harness one at a time. Remove the same-colored wire it replaces and immediately put the new wire on the open terminal, then connect the other end. Major appliances nearly always have a wiring diagram pasted somewhere on the inside of a panel or on their back that indicates the color coding of every wire and where it is connected, but it still helps to have the service manual available for reference. Replacing a wiring harness constitutes a major undertaking, and there are so many connections to be undone and remade that you need all the guidance you can get to complete the job satisfactorily.

SERIES AND PARALLEL CIRCUITS

As a footnote to the wiring found in appliances, it is useful to have a vague idea of what those wires are doing and how they are functioning. Most of the circuits in your home wiring system are arranged in what is known as *parallel* circuits (Fig. 175). Each fixture and receptacle has both a black hot line and a white return wire bringing electricity to and from it. Consequently, if one fixture or receptacle fails, the electricity in the branch line can still go to the other components and also return to its source of voltage. In other words, because each component has its own complete circuit, it can continue to operate so long as voltage is entering the branch line it is attached to.

Fig. 174. No question about it, a wiring harness can look rather confusing, but the wires are usually color-coded and often numbered as well.

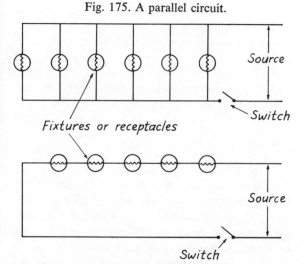

Fig. 175. A parallel circuit.

Fig. 176. A series circuit.

Series circuits are a little different, and are sometimes used in connecting components in appliances (Fig. 176). The series circuit has an unbroken return line, and all of its components are strung on the black line (the way switches are attached to house fixtures). As a result, when any of the components on the series circuits fails, it breaks the continuous path that electricity must have in order to flow, and all of the components on the circuit stop working because they stop receiving any power. The best example of series wiring is a string of Christmas tree lights, which works fine until one of the bulbs burns out and immediately all the bulbs on the line go dark.

In appliances, it is often desirable to have series wiring so that if one component fails, it shuts off the entire machine, or parts of the machine. The major advantage is that a group of components wired together in series can all be controlled by a single switch, so that parts of the appliance can be activated or deactivated at different times. It should be noted that the current interrupters attached to the branch circuits in your house are all connected to the branch lines in series and they function as a switch shutting down the entire branch circuit anytime there is a short circuit or overload.

Both series and parallel circuits are used in the wiring of home appliances. Sometimes they are applied separately, and sometimes they are used in combination (known as a series/parallel, or *combination* circuit), and there is usually no quick way of telling what kind of circuit you are dealing with. However, if you track the different wires and note what they are attached to, and whether there is an unbroken return line coming from the last in a series of components, you can usually discern the type of circuit you are working on.

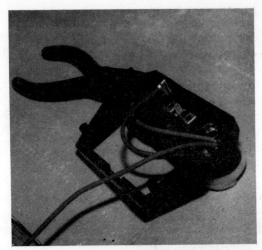

Fig. 177. Always grip the lug on a quick-connect terminal, never the wire.

flat prong that fits inside a correctly mated female lug crimped to the end of a wire. Be careful when removing a quick-connect connection to grip the *lug* with your fingers or pliers. Don't pull on the wire itself or it may come off its lug. Some expensive pieces of equipment, most notably compressors, must be replaced in their entirety if any of their quick-connect terminals are broken because the terminals are an integral part of the motor structure. Therefore, if you encounter a wire lug that is frozen to its terminal, cut the wire and splice a new wire to it, rather than risk damaging the terminal.

Self-locking terminals are little boxes with a hole in one side of them and a slot or hole next to that. The terminal accepts a solid wire, which it grips with a flat spring inside its box in the same manner as back-wired switches and receptacles. In order to free the wire, you must push a wire or awl into the slot and depress the spring while you pull the wire free.

TERMINALS

The wires in appliances may be fastened to several kinds of terminals. As mentioned before, the basic screw terminal is very prevalent and can receive either a bare wire or a round or U-shaped lug crimped on the end of the conductor.

Quick-connect terminals consist of a round or

HEATING ELEMENTS

Heat-producing appliances such as ovens, cooktops, toasters, hair dryers, irons, broilers, clothes dryers, and dishwashers have specially designed high-resistance wires (or rigid units) that resist electricity going through them so much that they become inordinately hot.

Fig. 178. You can see the coiled resistance wire in your toaster glow a bright red when the appliance is turned on.

The resistance wire can be a flat tape or coiled springlike configuration, and is made of metal alloys. The wire is typically supported in the appliance by mica holders, and of course to function it must form a complete circuit with its source of electricity.

Should a wire heating element break, you can make a temporary repair by connecting the ends with a crimp-on metal sleeve. Do not attempt to

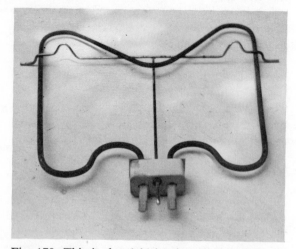

Fig. 179. This is the rigid heating element from an electric oven. It will also glow when electricity enters it via the plugs in the foreground.

splice the wire, since its loose ends may cause a short circuit. And as soon as possible replace the heating element. If you cannot do this with an exact replacement, at least use something that meets these standards: The length of the resistance wire must be exact; if the wire is coiled, the new wire must have the same number of coils per inch; the wire must also be able to meet the wattage and voltage ratings of the appliance. Take the old wire with you when you go shopping so you can come as close to duplicating it as possible.

Rigid heating elements are used in cooktops, ovens, stoves, frying pans, and dishwashers. They are nearly always in the form of a loop and have prongs at their ends which plug in to the side of the appliance. The rigid heating elements rarely burn out, but their replacement is as simple as unplugging the old unit and plugging in its replacement.

The one time you can destroy a rigid heating element is by washing it with oven cleaner. Many types of rigid heating elements react unfavorably if they are subjected to the acid in oven cleaners, which is one of the reasons the units are so easily removed. They are supposed to be taken out of the oven or cooktop before it is cleaned with powerful chemicals.

SWITCHES

By definition, a switch opens and closes contacts to complete an electrical circuit (Fig. 180). Switches have both mechanical and electrical elements, but in most cases when something goes wrong with a switch, the best remedy is to replace it with a unit having the same ampere/volt rating.

You can only determine whether a switch has failed mechanically by working it and observing whether it makes and breaks its contacts. Sometimes the contacts can be bent slightly so they will meet perfectly. They can also be cleaned with contact cleaner or very fine sandpaper, but if that does not solve the mechanical problem the switch usually must be replaced. Electrically, you can tighten terminal screws and clean their electrical contacts, but that is about all.

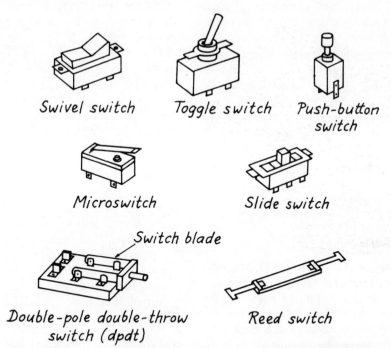

Swivel switch

Toggle switch

Push-button switch

Microswitch

Slide switch

Switch blade

Double-pole double-throw switch (dpdt)

Reed switch

Fig. 180. Some of the many types of switches used in appliances.

ROTARY SWITCHES

Rotary switches are found in large appliances having timers and consist of a dial attached to a shaft that holds a cam or series of cams. The cam is an irregular oval that depresses and releases a series of metal contacts as it rotates, opening and closing different circuits each time a contact is moved.

PUSH-BUTTON SWITCHES

Doorbells, chimes, and some small appliances use push buttons to depress a metal strip. The strip closes contacts to complete the circuit.

INFINITE-CONTROL SWITCHES

These are really rotary switches, often used in thermostats to control the temperature in ovens and ranges. Because the infinite control has no specific stopping points, it can be set anywhere along the distance of its rotation, so it is said to have an infinite number of control positions.

INFINITE HEAT-CONTROL SWITCHES

Another form of rotary switch used in toasters, broilers, and other small heating appliances incorporating thermostats is known as the infinite

Fig. 181. The front of an oven has some rotary switches to control the clock and timers. Directly below the clock is a push-button switch that controls the interior light, and the thermostat on the right side is, in effect, an infinite-control switch.

heat control. The control depresses a bimetal strip, which then contracts or expands to open or close the circuit.

A second infinite heat-control design involves a resistance wire wrapped around a bimetal strip. When the control knob is rotated, it moves a cam or a screw against the strip and current through the coil then causes the bimetal to bend so that it either opens or closes the circuit. With this type of device the harder the screw or cam presses against the bimetal, the farther away it is from its contact and the more heat is required to bend it enough to close the contacts.

SLIDE CONTROLS

Slide controls are flat versions of the infinite heat control. Instead of rotating a knob, there is a pointer or slide that is pushed back and forth along a bimetal track. The position of the slider regulates the distance between the bimetal and its contacts and varies the amount of heat (and therefore time) it takes the bimetal to bend toward or away from its stationary contacts.

SOLID-STATE CONTROLS

While generally speaking any switch that malfunctions is difficult to repair and usually must be replaced, solid-state controls absolutely cannot be repaired at home. A solid-state control consists of complicated circuitry printed on a mica-type board. More and more appliances are using them as a replacement for whole bags of other components, and it is safe to say that by the end of the 1980s practically all of the appliances manufactured will incorporate solid-state circuitry to some degree, if not totally.

Not much is ever supposed to happen to solid-state controls, but when it does the component is usually removed from the appliance and simply replaced. If you own a voltmeter you can check a solid-state switch in this manner:

1. Set the meter at its 150/200 AC scale.
2. Touch the meter probes to both terminals of the switch.
3. When the switch is closed, the meter should read between 1½ and 2 volts. If you get any other reading, the switch is faulty and should be replaced.

MICROSWITCHES

Microswitches are also appearing in more and more appliances, but they are almost always designed by the manufacturers to perform a specific chore in a specific unit. So, when one of them needs to be replaced, it is best to get its replacement from the manufacturer.

Microswitches are small boxes that are often shaped to fit into a particular area of the apparatus. For example, a microswitch might be used to shut off the circuit in a washing machine whenever the door is opened. There are likely to be one or more tiny buttons on the sides of the switch that are connected to a strip of spring metal inside its housing that opens and closes a circuit. The face of the switch is typically marked NO (Normally On) or NC (Normally Closed). When you replace a microswitch, be certain that your replacement is marked identically to the one you are removing from the unit.

Whenever the button(s) on a microswitch is depressed, you will hear a click from the metal strip inside the housing as it bends. If you do not hear the click, you can assume the switch is malfunctioning and should be replaced.

TIMERS

The core of any appliance that operates in cycles is its timer. Timers occur in dishwashers, clothes washers and dryers, ovens, and ranges. Their function is to move the appliance from one operation or cycle to the next, and then to shut off the machine when all the functions have been completed.

A timer consists of a series of cams attached to a shaft that usually has a selection dial or knob on its end. The cams are rotated by a small electric motor past a series of spring metal contacts. As each cam depresses the metal strips, the metal touches a stationary contact, closing a circuit. The cams do not rotate continuously, but actually jump ahead a few degrees every several seconds, to avoid any slow opening or closing of the circuits. The last jump in the cycle breaks the main circuit in the appliance and shuts off the entire machine.

While timers can look somewhat different from each other, they always have a number of

Fig. 182. Some timers are used to control lights and other appliances. The timer is plugged in to a receptacle and whatever it controls is plugged in to it.

The first area to look at is the timer motor. Turn on the appliance and observe the gears in the motor. They should all be rotating, although not at the same speed. If the motor gears are not turning properly, you may be able to unstick them with a few drops of oil, but more likely the motor must be disconnected from the timer and replaced. It will be bolted to the timer and have two wires leading from it to terminals on the timer housing. It should not require more than a few minutes to disengage.

If the motor is operating and the appliance still does not perform according to its cycle chart, the timer will most likely need to be replaced. The cycle chart is often pasted on the inside or back of the appliance cabinet, but it may also be reprinted in the owner's manual. You can check the timer by turning on the machine and following each of the cycles it goes through. The times for each cycle should match whatever the timing chart says they should be.

electrical wires connected to them (perhaps as many as two dozen or so) and they always have a synchronous motor attached to their side. Any time an appliance containing a timer begins to miss its cycles or in any way functions inappropriately, suspect the timer.

REPLACING A TIMER

Imposing as a timer is with its myriad of wires sticking out in all directions, replacing one is a remarkably simple procedure. If a timer on one of your appliances needs replacement, you must

Fig. 183. You can identify a timer immediately by all the wires attached to it, since it is the control center of whatever appliance it is in.

install an identical unit from the manufacturer. You could buy a new one, but they are expensive. At about a third the cost you can trade your broken timer in at an appliance parts store for a rebuilt unit. Rebuilt timers come with a one-year warranty, which is exactly what you get with a brand-new unit.

You do not yank your old timer out of the machine and cart it off to the parts store to use as barter. You go to the store and buy a rebuilt timer and give the store a small deposit, which you can retrieve when you have installed the replacement and taken the broken unit back to the store. There is a very good reason for doing business this way: It is considerably easier to transfer all of the wires between the timers by moving them one at a time. For best results, follow these directions:

1. Hold the replacement timer near the old unit and transfer each wire one at a time. The wires are all color-coded and may also have letters and numbers on them that correspond with similar identification printed next to each terminal on the timers. The terminals are usually quick-connect lugs. Between reading the numbers and letters and looking at the colors on the wires, it should take you about half an hour to get the twenty or so wires in a clothes-washer timer transferred from one unit to the other.

2. Unbolt the old timer when it has no more wires attached to it, and bolt the new unit in its place. Be sure that you position the replacement at exactly the same angle as the old unit.

3. Plug in the appliance and run it through all of its cycles. Using the owner's manual or the chart pasted on the machine, match each cycle with the amount of time it takes to complete.

If the new timer does not reproduce the correct times in each cycle, you have connected some of the wires to the wrong terminals. You will not have damaged the timer, you merely have to go back and make certain that the letters and numbers on each wire match the numbers and letters next to the terminal it is connected to.

THERMOSTATS

Thermostats control heat-producing or cooling devices, including ovens, ranges, clothes dryers, space heaters, toasters, broilers, and air conditioners. There are several types of thermostats ranging from relatively complicated gas-containing units down to simple bimetal strips and disks.

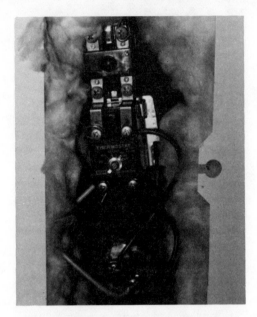

Fig. 184. Anybody can tell this is a thermostat because it says thermostat on it. In this instance the unit is buried in the insulation surrounding a hot-water heater and, like a timer, it controls what the appliance does or does not do.

BIMETAL STRIPS

Bimetal strips are widely used in heaters and toasters, and are connected to a stationary terminal at one end, but suspended *over* a stationary contact at the other end. The strip is composed of two layers of metal alloy, each of which expands and contracts at a different rate when it is heated or cooled. It is this different rate of change that causes the strip to bend toward or away from its contact.

When the bimetal strip is controlled by an infinite-control switch, the rotating knob on the switch drives a cam or screw against the bimetal, bringing it closer or farther away from its contact, and therefore requiring more or less heat to bend the strip far enough to open or close the contact.

THERMODISKS

Thermodisks show up in washing machines, dishwashers, some motors, clothes dryers, and ovens. They are concave or convex disks about an inch in diameter that are made of bimetal alloys and they open or close electrical contacts by "popping" when they are affected by the proper amount of heat.

All thermodisks and bimetal strips can malfunction; the best way to check them is to observe their action during operation of the appliance. If the strip or disk does not open or close its contacts, it must be replaced, which means you have to acquire a unit having an iden-

tical rating to the one you are replacing. Actual replacement is a matter of undoing a bolt or screw.

GAS-OPERATED THERMOSTATS

Used primarily in ovens and ranges, gas-operated thermostats have an expandable chamber, or bellows, filled with inert gas (Fig. 186). A thin tube is attached to the bellows and has a brass or copper bulb at its other end that is also filled with gas. The gas expands or contracts as it is heated or cools, causing the bellows to move toward or away from stationary electrical contacts. When the dial on the face of the thermostat is set to a particular heat, it alters the distance between the bellows and its contacts so that it must expand more or less before it can shut off the heat entering the oven.

Gas-operated thermostats must be repaired by specialists with special equipment (even professional repairmen send them out to be repaired or rebuilt). A rebuilt thermostat is approximately half the cost of a brand-new one and comes with the same one-year warranty, so if your thermo-

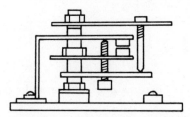

Fig. 185. Thermodisks and bimetal strips rarely need replacement. But if you have to change one, it is usually held in position by a screw or bolt.

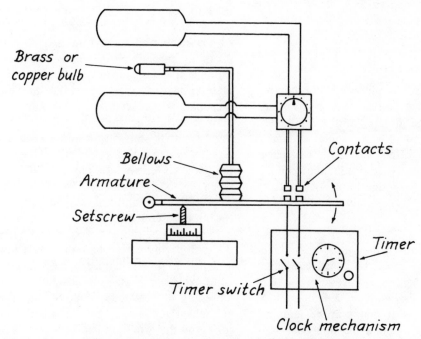

Fig. 186. Anatomy of a gas-operated thermostat.

stat develops a leak or breaks down, remove it and trade it in for a rebuilt one.

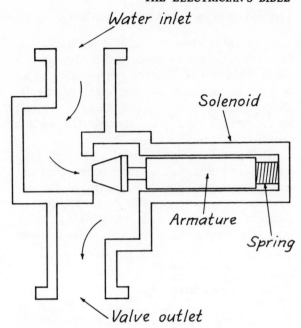

Fig. 187. Anatomy of a solenoid used to control a water outlet.

RHEOSTATS

Rheostats are used to control the amount of electricity entering an appliance and are made of a solid core wound with uninsulated resistance wire. A sliding contact moves along the resistance coil and picks up the flow of electricity at whatever point in the wire it happens to be touching. If the control point is near the beginning of the coil, it receives nearly all of the electricity flowing into the appliance. But if the control is at the far end of the coil, considerable amounts of electricity meet with resistance in the wire and reduced power is transmitted to the appliance, causing a lamp to dim or a motor to slow down.

Rheostats are gradually being replaced by solid-state switches in food blenders, mixers, and multispeed power tools. The solid-state controls generate less heat than a rheostat and are therefore considered more efficient.

SOLENOIDS

Solenoids are used to automatically operate switches, levers and valves in dishwashers, clothes washers, and dryers, and automobiles. The solenoid is a coil of wire wound around a hollow core (most often an insulated tube). When current is passed through the coil, it creates a magnetic field that pulls or repels a plunger through the center of the core (Fig. 187).

The terminals on a solenoid can corrode and need cleaning, especially if the unit is in conjunction with a valve or other water-using component. If the solenoid vibrates excessively, examine its plunger for misalignment, a bent or broken return spring, or dirt that may be jamming it. If the solenoid gives off a burning odor it is pulling too much current or is short-circuited and should be replaced.

RELAYS

Relays are electromagnetic switches commonly used to control motors, so you might encounter one in any motor-driven appliance. They consist of a wire wound around an iron core that produces a magnetic field whenever current passes through it. The magnetic field pulls at a metal plate known as its armature. The armature is attached to a contact that is closed for as long as current is flowing through the relay coil. Once the current is shut off, a spring pulls the armature away from the electrical contact, breaking the circuit. Relays are marked NO (Normally Open) or NC (Normally Closed), depending on whether the armature movement closes or opens the circuit it controls.

If the coil or the contacts on a relay burn out, the entire relay should be replaced. If the contact attached to the armature is misaligned with its mate, you can usually reposition the relay by loosening its setscrews or bolts and moving the unit slightly one way or the other. The return spring on the armature can be replaced by a similar spring found in hardware stores, but take the broken spring with you when you buy a replacement so you can duplicate it as closely as possible.

SENSORS AND RESPONDERS

Sensors and responders are always used together, principally in clothes dryers and ovens. The sensor is a tiny disk or tube that contains a pair of electrodes that react to heat by transmitting electricity to the responder. The responder is a little piece of plastic containing several terminals that connect wires leading to the switch that the responder/sensor combination controls. Electricity coming from the sensor is transmitted to the switch via the responder to open or close the switch and turn the heat on or off.

If a sensor fails to operate, the appliance will stop working; if a responder fails there will be too much heat in the appliance, or the heat will come on slowly or not at all. About the only repair you can make is to clean and tighten the contacts on the responder and the switch it is connected to. Otherwise, the sensor and responders must be replaced with identical replacements procured from the appliance manufacturer.

PUMPS

The kind of centrifugal pumps found in dishwashers and clothes washers have both intake and outlet ports on opposite sides of their housing that are connected to water hoses. In between the ports is a multibladed fan, or impeller, which is connected to the shaft of a motor, either directly or via a belt and pulley system. When water enters the intake port the rotation of the impeller driven by the motor thrusts it out the outlet port.

Uncomplicated as it is, several maladies can affect a pump. The clamps holding the intake and outlet hoses may loosen, causing water, or whatever fluid is going through the pump, to leak. The hose clamps can usually be retightened. The bearings on the impeller shaft must be lubricated periodically to prevent them from

Fig. 188. A pump is normally just a multibladed fan attached to the shaft of a motor. It must be sealed in its housing in order to maintain enough pressure to move the water through it.

being seized, or "frozen." If the bearings seize, a thorough cleaning and lubrication may (or may not) free them. If it doesn't work, the bearings will have to be replaced.

There are also gaskets and seals between all the parts that make up the housing to render the unit watertight. These can crack and dry out, at which point they will allow leakage and must be replaced. If the housing itself becomes so corroded that it is affecting the operation of the pump, replace the entire unit.

CHAPTER NINE

Motors

Motors are among the devices that can be used to convert electrical energy into mechanical work. As such, they have numerous applications both in industry and around the home, and can be found in a full range of sizes and amounts of power delivered. In and around your house, most if not all of the motors are fractional horsepower devices used to rotate the washing or drying drums in your clothes washers and dryers, or power the pump in your dishwasher, or run most of your electrically powered shop tools.

Electrical motors are all rated according to their horsepower (or partial horsepower). There is usually a plate attached to the housing of the motor that provides several bits of information, including the machine's voltage, amperage, revolutions per minute (RPM), cycles (Hertz), phase (1 or 3), and horsepower. A horsepower is defined as the work needed to lift 33,000 pounds one foot, which is equal to lifting 550 pounds one foot in one second. Translated into electrical terms, 1 horsepower equals 746 watts. Note, however, that in order for a motor to deliver 1 horsepower (746 watts), it will actually consume almost 1000 watts from its power source. The lost 254 watts is dissipated as heat from resistance in the motor windings, friction from the motor bearings, and other such factors.

SOME THEORY BEHIND GENERATORS AND MOTORS

How both generators and motors function can be demonstrated by manually rotating a single loop of wire between the north and south poles of a permanent magnet (Fig. 189). The movement of the wire generates what is called induced AC voltage. The loop has no north or south poles, but if current flows through the wire it will travel along opposite sides of the loop in opposite directions. This causes the magnetic fields on opposite sides of the wire to be in opposition, so if you put the loop inside the magnetic field between two magnets, you have electromagnetic force traveling up one side of the loop and at the same time going down the opposite side. And what that does is make the loop rotate.

Because the magnetic lines of force are moving in opposite directions on opposite sides of the loop, the voltage produced by a rotating armature arrives at the external circuit attached to a generator as an alternating current. The wire keeps rotating in the same direction, but the current comes blasting out of the ends of the wire at

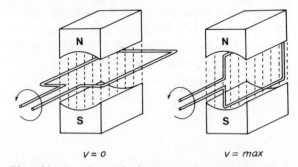

V = 0   V = max

Fig. 189. As a loop of wire rotates between the poles of a magnet, it produces alternating current. When the loop is parallel to the magnetic field the result is zero voltage (a). As the loop rotates to cut across the field, the induced voltage reaches its maximum output (b).

186,000 miles per second as it goes in first one direction and then the opposite.

Generators and motors both consist of two basic parts, an armature or *rotor,* and field windings or *stator* (Fig. 190). The armature is made up of a series of wires wound around a metal core. The core has a shaft through its center so that it can rotate, and the wire loops around the core are attached to a circuit. In generators, which are powered by some form of fuel, the wire loops around the armature are connected to a circuit that carries the output voltage from the generator as long as the armature rotates inside the field winding.

The field winding is also a series of coils, but they are wound around a hollow metal core so they form an electromagnet that provides a magnetic field for the armature to rotate within.

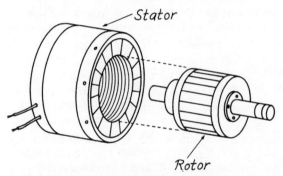

Fig. 190. Motors and generators consist of a stator and a rotor which rotates through the center of the wires wound around the stator core.

AC INDUCTION MOTORS

AC induction motors have an armature and a field winding, which is connected to a source of alternating current. While the immediate source of power may be a receptacle in your home, ultimately it is a generator belonging to your utility company.

The stator (field windings) induces alternating current to the rotor without physically touching it by way of the magnetic field that is established whenever electricity flows through the windings. It is this magnetic field that reacts with the stator field that causes the rotor to spin. Because the amount of rotation is directly proportional to the amount of current flowing through the coils, the rate of rotation can be increased or decreased

simply by changing the amount of current that passes through the field windings.

The rotor, as it spins inside the center of the field windings, is attached to whatever gears, pulleys, or other mechanical devices are needed to convert its motion into mechanical work.

SHADED-POLE MOTORS

The simplest form of AC induction motor is known as the shaded-pole motor. Instead of having several windings around its stator, the wires are all wrapped around only one side of it. And rather than windings on the rotor, there are sets of shading poles in the form of thick copper bars or wires. In spite of their design, shaded-pole motors function the same way any other motor does, by creating magnetic fields that force the rotor to rotate.

Shaded-pole motors are normally small and produce a low torque, so they are used only in small motor-driven appliances such as fans, electric toothbrushes, can openers, some hair dryers, and the rotisseries attached to broilers and ovens. They have proved themselves as pretty much trouble-free, and usually when they do malfunction the best repair is to replace the entire motor.

SHADED-POLE MOTOR REPAIR CHART

PROBLEM: Motor doesn't start.

CAUSES	REPAIR
No power.	Replace fuse or reset circuit breaker.
Plug unseated.	Insert plug in receptacle.
Cordset bad.	Check for loose terminals and broken or worn wires. Clean and tighten terminals; replace or repair wires.
Faulty switch.	Clean and tighten switch terminals; replace switch.
Seized (frozen) bearings.	Lubricate bearings; replace if worn.
Shaft does not turn.	Clean shaft; if bent or worn, replace shaft.
Open field windings.	Test for continuity. If no continuity, replace the coil or get a new motor.

PROBLEM: Motor overheats.

CAUSES	REPAIR
Overloaded circuit.	Reduce the load on the motor and restart it.
Seized bearings.	Lubricate or replace bearings.
Short circuit in field windings.	If no continuity in the field windings, replace the coil or the entire motor.

PROBLEM: Motor has poor torque.

CAUSES	REPAIR
Insufficient power.	Replace line cord.
Short circuit in field windings.	Replace field windings or the whole motor.
Seized bearings.	Lubricate; replace bearings.
Worn shaft.	Clean shaft; replace if bent or badly worn.

PROBLEM: Excessive vibration or noise.

CAUSES	REPAIR
Worn bearings.	Lubricate or replace.
Shaft worn.	Clean; replace if badly worn.
Loose parts.	Tighten all loose parts; replace any worn or broken components.

PROBLEM: Smokes.

CAUSES	REPAIR
Shorted field coil.	Replace coil or entire motor.
Seized bearings.	Lubricate; replace if worn.

SPLIT-PHASE MOTORS

Split-phase motors have two sets of windings (which constitute two circuits) around their stators, known as the *main running,* and the *starting* windings. When the motor is first started both circuits are in use until the rotor reaches about three quarters of its normal running speed. At that point the added current from the starting winding is no longer needed and is automatically switched off.

Split-phase motors deliver considerable torque and are therefore relatively powerful. So they are used to run compressors, washing machines,

Fig. 191. A split-phase motor. In this instance its stator windings are covered in insulating paper.

clothes dryers, dishwashers, and most stationary tools other than saws (grinders, drill presses, lathes, sanders, joint-planers).

SPLIT-PHASE MOTOR REPAIR CHART

PROBLEM: Does not run.

CAUSES	REPAIR
No power.	Reactivate current interrupter. Reset motor overload button. Tighten loose wire terminals. Reseat plug in receptacle.
Faulty cordset.	Repair or replace worn wires and loose terminals.
Faulty switch.	Clean and tighten contacts; replace switch if no continuity.
Seized bearings.	Lubricate; replace if necessary.
Field shorted.	Replace field windings or entire motor.
Open circuit in windings.	Replace windings or entire motor.
Shaft does not turn.	Clean and lubricate shaft; replace if badly worn or bent.

PROBLEM: Overheats, does not attain running speed.

CAUSES	REPAIR
Motor overloaded.	Remove belt. If motor runs, reduce its load.
Field shorted.	Replace field windings or entire motor.

CAUSES	REPAIR
Windings shorted.	Replace windings or entire motor.
Seized bearings.	Lubricate or replace bearings.
Faulty switch.	Clean contacts; if parts are burned or broken, replace switch.

PROBLEM: Motor vibrates or is overly noisy.

CAUSES	REPAIR
Worn bearings.	Lubricate or replace bearings.
Loose parts.	Clean and tighten all parts.
Pulleys out of alignment.	Realign or replace pulleys.
Worn belt.	Replace belt.
Shaft seized.	If worn or bent, replace.

PROBLEM: Bearings wear too much.

CAUSES	REPAIR
Belt too tight.	Loosen belt; adjust as per owner's manual.
Dirty oil.	Clean bearings and relubricate.
Dirty bearings.	Clean; if worn, replace bearings.
Windings shorted.	Replace windings or entire motor.
Switch faulty.	Clean and tighten contacts or replace switch.
Seized bearings.	Lubricate; replace worn bearings.

CAPACITOR-START MOTORS

Appliances such as air conditioners and refrigerators, stationary saws, or any appliance needing a motor with a high-starting force (torque), employ a *capacitor-start* motor. The motors themselves are a split-phase type, but with a capacitor added to provide extra power to the starting windings.

The capacitor is a storer of electrical charge. It is capable of building up a high voltage, which it then releases in short bursts to assist the motor as it is starting up. Physically, the capacitor is a round or square canister usually housed in a

Fig. 192. Capacitor-start motors are split-phase motors that need a capacitor to help get them started. One of the lead wires coming from the motor is attached to the capacitor, along with one wire of the motor's power cord.

metal cover attached to the side of the motor housing. The canister contains chemical-impregnated paper which allows it to retain voltage for relatively long periods of time. While the repair of capacitor-start motors is no different than any split-phase motor, the capacitor demands a certain amount of special attention.

Fig. 193. Capacitors can be round or square and are used to store electricity.

TESTING CAPACITORS

Any time you intend to work on a capacitor, first disconnect the motor. Then discharge the capacitor by placing the blade of a screwdriver against its terminals. You will get some sparks, but the capacitor will discharge its stored voltage.

If the motor hums but does not start up, or starts very slowly, the odds are the capacitor has lost its oomph and needs to be replaced. If you have an ohmmeter you can test the capacitor in this manner:

1. Disconnect the motor from its power source.

2. Discharge the capacitor.

Fig. 194. Pull the quick-connect terminal lugs off the capacitor. Be careful not to touch the terminals.

3. Set the ohmmeter to a range that has its highest ohmage above the resistance rating printed on the side of the capacitor.

4. Touch one probe of the meter to each of the terminals on the capacitor. The meter should read a low ohmage and then take several minutes to move up the meter scale to the capacitor's rating. If the needle stays at the low end of the meter scale, the capacitor is short-circuited. If the meter needle maintains a steady high-ohms reading, there is an open circuit in the capacitor. In either case the capacitor cannot be repaired; it must be replaced. Disconnect its terminals, remove the unit from its housing, and install a new capacitor having the identical rating as the old unit.

Fig. 195. Place the blade of a screwdriver against both terminals at the same time. There will be a scary little spark as the unit short-circuits and drains off its charge.

CAPACITOR-START MOTOR REPAIR CHART

PROBLEM: Does not start.

CAUSES	REPAIR
No power.	Replace the fuse or reset the circuit breaker. Reset motor overload device. Tighten loose parts. Be sure the plug is seated in its outlet.
Cordset faulty.	Repair or replace cordset.
Seized bearings.	Lubricate; replace worn bearings.
Switch faulty.	Clean and tighten terminals; replace switch.
Open circuit in starting windings.	Replace windings or entire motor.
Faulty capacitor.	Replace.

PROBLEM: Motor hums; does not operate.

CAUSES	REPAIR
Faulty capacitor.	Replace.
Open in starting windings.	Replace windings or entire motor.
External overload.	Disconnect belt. If motor runs, reduce load on the motor.

PROBLEM: Smokes.

CAUSES	REPAIR
Short in windings.	Replace windings or entire motor.
Faulty switch.	Clean and tighten terminals; replace switch.
Seized bearings.	Lubricate; replace bearings.
Motor wet.	Let the motor dry completely before using.

UNIVERSAL MOTORS

The universal motor is powerful, versatile, and can be run on either AC or DC, as well as be made to reverse the direction of its rotation.

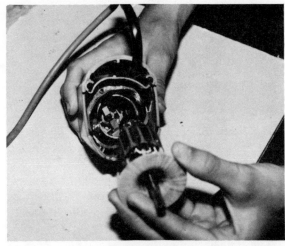

Fig. 196. Universal motors consist of a stator and an armature, and behind that a commutator.

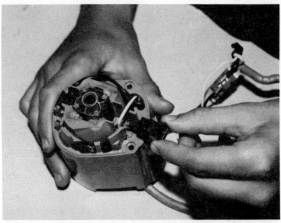

Fig. 198. The brushes are held against the commutator by springs. There will be a brush on opposite sides of the commutator. When one brush needs replacing, change both of them.

Consequently, universal motors are employed in any appliance or tool requiring medium power and a variety of motor capabilities. Mixers, blenders, sewing machines, vacuum cleaners, and such hand power tools as drills, saws, and sanders all are likely to incorporate universal motors. Moreover, all battery-powered appliances must also have a universal motor, since batteries provide only direct current. In other words, your hedge clippers, portable tools, and every other battery-powered, motor-run appliance all have universal motors.

against the spinning commutator by springs on opposite sides of the commutator. Therefore the brushes constantly make electrical contact with the brass bars on the commutator. The commutator converts AC power to DC by reversing the loop connections to the brushes every half cycle, causing a constant polarity of voltage coming through the brushes from the source of power.

UNIVERSAL MOTOR REPAIR CHART

PROBLEM: Does not start.

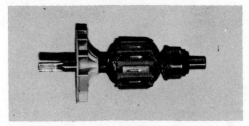

Fig. 197. From left to right is the shaft, the fan, the rotor, the commutator, and the other end of the shaft.

Like all motors, a universal motor has a stator and a rotor, but it also has a *commutator* and *brushes*. The commutator is attached to the rotor and consists of copper bars separated from each other by mica strips. Each pair of bars is connected to one loop in the rotor, and the entire commutator rotates between two stationary brushes. The brushes are made of compressed graphite to provide a low resistance, and are held

CAUSES	REPAIR
No power.	Reset the current interrupter. Clean and tighten loose terminals. Be sure plug is seated in its receptacle.
Faulty cordset.	Repair or replace cordset.
Faulty switch.	Repair or replace the switch.
Worn brushes.	Replace both brushes.
Brushes sticking.	Clean brushes; replace if worn.
Seized bearings.	Lubricate; replace if worn.
Armature does not rotate.	Clean armature shaft; replace if worn.
Faulty speed control.	Replace.
Armature windings open or shorted.	Replace windings or entire armature.

PROBLEM: Motor vibrates or is too noisy.

CAUSES	REPAIR
Worn bearings.	Lubricate; replace.
Armature does not turn.	Clean shaft; replace if badly worn.
Loose parts.	Clean and tighten all loose parts.

PROBLEM: Overheats; smokes.

CAUSES	REPAIR
Armature windings shorted.	Replace armature.
Field windings shorted.	Replace coil or entire motor.
Seized bearings.	Lubricate; replace if worn.
External overload.	Reduce load.
Motor wet.	Allow to dry before using.

PROBLEM: Sparks.

CAUSES	REPAIR
Short in armature coils.	Replace armature.
Field coils shorted.	Replace motor.
Brushes worn or out of alignment.	Realign or replace brushes.
Commutator pitted or damaged.	File down mica strips; replace commutator if bars are loose.

SYNCHRONOUS MOTORS

Synchronous motors are small, specially engineered motors used in electric clocks and timers. They are designed to maintain a rotation speed that is synchronized with the cycles of alternating current (60 alternations per second). Most synchronous motors can only be repaired with a set of jeweler's tools, and in any case they are inexpensive enough so that practically anybody can afford to buy a replacement if the motor breaks down.

REPAIRING MOTORS

A motor that has stopped running may need lubrication or new bearings, its switches may have become dysfunctional, or its windings may have burned. When troubleshooting any motor, this is the procedure that can be followed:

1. Disconnect the motor from its source of power.

2. Check the cord and plug for any bared wires, loose terminals, or dirty contacts and make the necessary repairs.

3. If the motor is a universal motor, unscrew the brush holders and examine the brush springs for wear or breaks; the brushes should be longer than they are wide. If one brush is worn down too far, replace both brushes.

New brushes have squared ends which must be shaped to fit the curve of the commutator. Wrap a strip of sandpaper around the commutator and push the brushes against it as you rotate the sandpaper. When the brushes fit the contour of the commutator, remove the sandpaper. The brushes should move freely in their holders and their springs must hold them firmly against the brass bars.

4. Wiggle the motor shaft. It should not jiggle at all. Rotate the shaft. It should spin freely. If the shaft jiggles, the bearings are probably worn and should be replaced. If the shaft does not allow the armature to spin freely, it may be bent or the bearings may be worn or seized. Your chances of straightening a bent shaft are minimal; replace it.

5. Plug in the motor and turn it on. Touch one probe of a hot-line tester to the running winding (the larger gauge wire) and the other probe to the core of the stator (the laminated metal plates). If the neon test lamp goes on, the winding is grounded and there is a wire somewhere that is touching either the rotor or the stator cores.

6. Now run the motor for a few seconds. If it smokes, makes excessive noise, seems to run too slowly or too fast, or if it blows the circuit interrupter, one of its windings is either short-circuited or partially burned out. You can have the windings rewound (it requires special machinery), but it is usually less expensive to buy a new motor.

7. Take off the plates on both ends of the

motor and withdraw the armature from its stator. If the wires on either the stator or the rotor
are black, they are burned and must be replaced.
Or buy a new motor.

TESTING MOTORS

If you have a voltohm meter there are some
tests you can make on your motors:

LOOKING FOR AN OPEN CIRCUIT

1. Set your meter to its Rx100 scale.
2. Touch each meter probe to a lead from the
field winding (on the stator). If the meter reads
infinite ohms, there is a broken wire in the coil.
3. If the motor is a variable speed machine,
there will be several leads coming from the stator. Test each of them.

LOOKING FOR CURRENT LEAKAGE

1. Set your meter to its Rx100 scale.
2. Touch one meter probe to any lead coming
from the winding in the stator.
3. Touch the other meter probe to the motor
frame. If the meter reads high, current is leaking
to the motor frame (the motor is probably giving
you a shock every time you touch it when it is
running). There is no current leakage if the
meter reads infinite ohms.

If you are testing a universal motor, you can
perform the above test on the armature (rotor)
by touching one meter probe to the motor shaft
and the other to each commutator bar in turn.

TESTING COMMUTATORS (universal motors only)

There is a short circuit in the commutator if
the motor runs hot, sparks, or will not run at all.
Your problem is to find out where the short circuit is.

1. Set your voltohm meter to its Rx1 scale.
2. Touch the meter probes to any two adjacent bars on the commutator. Write down the
meter reading.
3. Touch the meter probes to the next two adjacent bars on the commutator and note the
meter reading. Continue testing adjacent bars on

the commutator until you have a reading for
each pair of bars.

If any one reading is considerably lower than
the others, there is a short circuit between those
two bars. If any reading is unusually higher than
the other readings, the coil in the armature between those two bars is broken. With either situation, the commutator must be replaced and you
may find it cheaper and easier to get a whole
new motor.

OTHER UNIVERSAL MOTOR TESTS

1. Universal motors always have a fan attached to the end of their shafts. The fan must
be secure on the shaft and none of its blades
should be bent or broken. If the blades are bent
you may be able to straighten them out. If they
are broken, replace the fan.
2. The brass bars on the commutator are separated by insulating mica strips that should not
extend above the top of the bars. If any of the
mica strips are too high, shave them off with a
knife until they reside below the surface of the
bars.
3. If the brass bars on a commutator are pit-
ted, discolored, or overly shiny, there is a short
circuit in the armature (rotor) coil. If the bars
are rough, the brushes are defective and should
be replaced. You can remove the roughness by
sanding the bars with fine sandpaper.

MOTOR CONTROLS

There are a variety of devices used to control
the speeds of motors. Most of them function by
varying the magnetism created by the windings
in the motor.

RHEOSTATS

As previously described, rheostats are a solid
metal core wound with coils of uninsulated wire.
A sliding contact can be moved along the resistance coil to increase or decrease the amount of
current entering the motor and therefore increasing or decreasing the magnetism in the motor
windings, which in turn speeds up or slows down
the motor. Rheostats are often used on the
motors in blenders and mixers.

TAPPED FIELD

Tapped field controls are also used in blenders, mixers, and variable-speed power tools. By rotating the control knob, the tapped field control touches the stator windings at different points to provide more or less current in the winding. The more wire that the tapped field device uses, the greater its resistance and the slower the motor will rotate.

DIODES (RECTIFIERS)

Diodes allow current to pass through them in only one direction. Since AC alternates 120 times a second, a diode installed in an AC line will reduce its voltage by one half, because only half the current will be going at the right direction to get through the diode.

Diodes are solid-state devices and are very delicate. They are also soldered in place. In order to test a diode you must remove one end of it from the circuit so that your meter readings will not be affected by other components on the circuit.

1. Touch one of the soldered connections of the diode with the tip of a soldering iron until the solder melts enough to free the connection.

2. Set your ohmmeter at its Rx100 scale.

3. Touch the meter probes to the terminals of the diode and note the reading.

4. Reverse the meter probes on the diode terminals and note the reading. One reading should be less than 100 ohms and the other reading should be more than 1000 ohms. If the readings are identical, the diode is defective and must be replaced.

When you install a new diode, be sure you put it on the circuit aimed in exactly the same direction as the unit you are replacing; remember that diodes only permit current to pass through them in one direction.

Diodes are used on appliances having 2-speed motors (full speed and half speed), such as 2-speed drills, mixers, and hair blowers and dryers.

CENTRIFUGAL GOVERNORS

Centrifugal governors are mounted on the shaft of the motor they control, and the motors are almost always split-phase. The governor has a stationary component having a pair of contacts which, when the motor is not running, are held together by a spring in the rotating (centrifugal) part of the switch. When the motor is started, the switch completes the contact between the starting winding and the source of power, so that both the main and the starting windings are receiving current to get the motor running. As the motor approaches its running speed, the switch also rotates, pulling away from its contacts and removing the starting winding from the circuit.

ELECTRICITY SAVING CHECKLIST

Electricity continues to rise in cost. A few years ago you paid less than 3 or 4 cents a kilowatt hour. Now the price is several times that and rising, so that it is not uncommon for a family having a normal complement of modern appliances to pay close to $100 a month to their local electric company.

There are some things you can do to save electricity and keep your electrical bill down to manageable proportions. Some of these energy-saving steps can be done once and have the continuous effect of saving you money. Most of them, however, constitute an ongoing program and developing the habit of conservation.

REFRIGERATOR/FREEZER MANAGEMENT

1. Don't open refrigerator and/or freezer doors more than you have to, especially during hot weather. Plan ahead and take out as many items as you will need at one time.

2. Don't set a freezing unit's temperature too low. Try different temperatures until you find the one that will keep ice cream solid but not hard as a rock. Refrigerators should be run at between 34° and 37° F.

3. Keep freezers full. Frozen foods tend to assist the machine in remaining cold.

4. Try never to load your freezer with huge amounts of unfrozen food at the same time. The machinery will have to work overtime to get it all down to freezing temperatures.

5. Don't stuff refrigerators so full that no cold air can circulate around the food.

6. Allow hot dishes time to cool before you

put them in a refrigerator; their warmth will force the machine to run inordinately long.

7. When you go away from home more than three days, empty the refrigerator as much as you can and raise its temperature setting a few degrees.

8. Keep the refrigerator and freezer clean. Vacuum the condenser coils about every three months.

9. If you have a frost-free machine, keep any liquids in it tightly covered, so they will not evaporate. Evaporation causes frost-free units to work overtime.

10. Never install any freezer or refrigerator near heat-producing appliances. Always allow for air to circulate behind the unit, which means don't bury the back of it in a wall.

DISHWASHER MANAGEMENT

The latest dishwashers have a number of energy-saving capabilities, many of which you should consider if you are buying a new appliance. Even with the new dishwashers, following these steps will reduce their electrical demands:

1. Only run the dishwasher when it is full.

2. Scrape away food from all of the dishes to be washed and skip the pre-rinse cycle.

3. Load the machine properly, as per the manufacturer's suggestions stated in the owner's manual.

4. Keep the filter screen free of food particles.

5. Shut off the dishwasher when it reaches its drying cycle (newer units often have a control position to do this), then open the door and let the dishes air-dry. You can't imagine how much electricity you will save.

6. Use the right amount of detergent. Too much, too little, or the wrong kind of detergent reduces the efficiency of the appliance.

OVENS AND STOVES

Next to hot-water heaters, these use more electricity than any other appliance you own, but there are a lot of ways to use them without spending more than you have to for energy.

1. Small amounts of water heat faster than large amounts. Use the smallest amounts of water necessary for whatever you are cooking.

2. You can save 20 percent of the energy you use just by putting a lid on pots and pans; water boils faster when it is covered.

3. When the liquids you are boiling reach the boiling point, cut back the heat as far as you can while still keeping the contents bubbling.

4. Fit the pot to the burner. Boiling a small pot of water on a large burner element uses more electricity than necessary.

5. Thaw foods before cooking them.

6. Keep the bottoms of your pans clean. The shinier they are, the more efficiently they will heat.

7. Whenever possible, use stove-top burners to cook fast foods and the oven for foods that require a long cooking time. The stove-top units consume less energy than the oven, but ovens retain their heat longer.

8. Microwave ovens are very energy-efficient. Use one whenever possible.

9. Don't turn on a cooking element until the pan is on it.

10. Use the type of cookware in your stove recommended by the appliance manufacturer. In general, copper and stainless steel need less electricity than aluminum.

11. When you use your oven or broiler, cook as many dishes as you can at the same time.

12. Most foods do not really need a preheated oven. If you absolutely have to preheat, get the food in it as soon as the cavity reaches the desired temperature. Never preheat a broiler.

13. You can shut off an electric burner five minutes before the food is cooked and you will still get enough heat to finish your cooking.

14. When baking with ceramic or glass dishes, you can almost always use lower temperature settings than the recipe calls for.

15. Never waste electricity trying to heat your kitchen with an oven or stove.

16. Don't open an oven door more than you have to during a cooking process. Every time you open the oven door you lose mountains of heat.

17. If you have a double oven, always use the smaller cavity whenever possible. It uses less electricity.

18. Use your smaller cooking appliances

(toasters, tabletop broilers, Crockpots, electric frypans) whenever possible instead of your oven or stove. They consume less electricity.

19. If you have a self-cleaning oven, it operates in its cleaning mode at about 900° F. So clean the oven immediately after you have used it for cooking, when the appliance is already halfway to the cleaning temperature.

CLOTHES WASHER MAINTENANCE

Clothes washers use both electricity and water, and can easily waste both resources, particularly if the machine is not in proper working order.

1. If you can control the water level in your machine, keep it to its lowest level.

2. Don't overload the machine, or it will have to work beyond its normal capacity.

3. Clean the lint filter after every washing.

4. Use the coldest water possible, whenever possible. Hot water, particularly if you have an electric water heater, costs electricity.

5. Wash similar fabrics together and don't use more detergent than is necessary. Oversudsing overworks the machine.

CLOTHES DRYER MAINTENANCE

1. Run your dryer only when you have a full load.

2. Set the dryer timer to the shortest period of time that will dry whatever you put in it. Overdrying costs money.

3. If you can, dry similar fabrics together, and use consecutive loads. Dryers hold their heat for some time, and you can use that heat by going from one load to the next.

4. Clean the lint filter after each load.

5. If you plan on ironing your clothes, don't dry them completely. You are only going to dampen them later anyway.

HOT-WATER HEATER MAINTENANCE

You could go through a lifetime and never even notice the electric hot-water heater in your basement. It has no moving parts and makes no noise as it obediently turns on and off all day and all night, to maintain its 50–100 gallons of water hot and ready for use. Next to your furnace, it is consuming more energy than any other electrically operated machine in your home.

1. Feel the outside of the heater. If it is warm to your touch, it is losing heat through its interior insulation and should be wrapped in blanket insulation. The blanket is merely wrapped around the outside of the unit and taped together.

2. Hot water loses one degree of heat for every foot it travels through a copper pipe. You can prevent the water from getting cold (and therefore having to be replaced by the heater at considerable cost) by wrapping all exposed hot-water pipes with insulation.

3. Set the thermostat(s) on your heater to the minimum prescribed setting. If you have a dishwasher, the heater temperature setting must be at 140° F. Otherwise, 120° is sufficient.

4. Install a timer on the heater cable. The timer can be used to shut off the heater at night (few people require hot water during the hours they sleep) and also during times of the day when you are away from home. You could probably reduce your electrical consumption by 20 percent with a properly set timer attached to your electric hot-water heater.

5. Turn off the heater altogether if you plan to be away from home for more than a day.

AIR CONDITIONER MAINTENANCE

The ways of saving electricity with your air conditioner apply to room-sized units as well as central A/C systems.

1. Keep the air vents in the conditioner aimed at the floor, which is where cold air naturally settles anyway.

2. Close all heating system vents in the rooms you are cooling so you don't waste cold air going down the vents, and air-conditioning your furnace, which is not running anyway.

3. Use shades, drapes, blinds, and closed windows to keep direct sunlight away from the air conditioner.

4. Turn off all lights as much as possible. Bulbs throw off tremendous amounts of heat, which will make your A/C work harder.

5. Shade the exterior side of your A/C unit and be sure it is not blocked from air circulation around it.

6. Do not block the vents in the machine with drapes or furniture.

7. Turn on exhaust vents to remove moisture in kitchens and bathrooms, but shut them down as soon as the moisture has disappeared.

8. If you are out of the house every day, install a timer to keep the A/C off until about an hour before you plan to be home.

9. Turn off your central A/C if you are going to be away from home for more than a day.

10. Keep the air conditioner, whatever type it is, in proper working order and, above all, clean its filters regularly.

LIGHTING

Lights consume surprising amounts of electricity, so you need to be very careful about how they are used if you are at all concerned about the cost of electricity.

1. Turn off incandescent lights every time you leave a room.

2. Leave fluorescent lights on when you leave a room if you plan to turn them on again within 15 minutes. Fluorescents demand considerable amounts of power to get started and very little to keep running.

3. Wherever practical, convert your incandescent lamps to fluorescents. Fluorescents are ideal over work spaces and in hallways.

4. Install dimmer switches. They help you use less electricity and also preserve the lifetime of your bulbs.

5. Use 3-way bulbs whenever possible so that you can adjust the lighting intensity to different lighting needs.

6. Darkened bulbs give off less light and should be replaced.

7. Keep your bulbs clean so they will give you maximum light at all times.

8. Where fill-in or general lighting is needed, such as in a hallway, use the smallest wattage bulb that will light the area.

9. Install photoelectric cells or timers to outdoor security lighting.

10. Use your windows as a light source as much as possible. In other words, don't turn on a light until it is absolutely necessary.

11. Be sure all bulbs in remote places (attics, basements, garages, sheds) are not left burning.

12. Use spotlights over work areas where a specific task is carried on. Use reflective bulbs and fixtures to intensify light wherever it is needed.

13. Use light colors to help reflect available light.